RADIATION

AND MODERN LIFE

Fulfilling MARIE CURIE'S Dream

Advance Praise for *Radiation and Modern Life*

"This book is a treasure trove of information about the uses of radiation and radioactive materials told in a most interesting and meaningful way. It is easy reading and communicates the basic principles of radioactivity and the significance of each application of radiation in a personalized way that is very understandable, even by readers without a background in science."

—Robert J. Lull, MD,
Clinical Professor of Radiology and Laboratory Medicine and Director,
Nuclear Medicine Resident Training Program,
University of California–San Francisco School of Medicine;
Chief of Nuclear Medicine, San Francisco General Hospital;
Past President, American College of Nuclear Physicians

"This exceptional book dispels many false notions and adds a wealth of wonderful information about how our lives are and can be better through the wise use of radiation technologies. Properly applied in food processing, radiation can add up to 25 percent to the world's pure food supply, thereby saving millions upon million of people from starvation and disease."

—Don Hodel,
Former Secretary of Energy and Secretary of the Interior

"Alan Waltar has rendered a great public service in compiling and explaining in easily understandable terms the diverse vital contributions of nuclear technology to our economy and to our way of life. His thorough research in digging out applications in numerous very different areas is truly remarkable, and he tells the story in a way that readily maintains the interest of the reader. Anyone, scientist or nonscientist, will learn a great deal from this book."

—Bernard L. Cohen,
Professor Emeritus, Department of Physics, University of Pittsburgh,
and author of *The Nuclear Energy Option*

"Dr. Waltar reveals in this book the startling fact that many areas of our everyday life have been beneficially affected by harnessed radiation, helping the readers to take the mystery and fear out of radiation, which is as much a part of our natural surroundings as air and water."

—Shunsuke Kondo,
Chairman of the Japan Atomic Energy Commission,
Professor Emeritus of the University of Tokyo

"Waltar presents the public with important technical insights on the multiuses of radiation in our daily lives in a relevant, accessible manner. His discussion 'How Harmful Is Ionizing Radiation' is a wonderfully clear and convincing dialogue."

—Neil E. Todreas,
KEPCO Professor of Nuclear Engineering,
Massachusetts Institute of Technology

RADIATION
AND MODERN LIFE

Fulfilling MARIE CURIE'S Dream

ALAN E. WALTAR

Introduction by
DR. HÉLÈNE LANGEVIN-JOLIOT,
Granddaughter of Marie Curie

59 John Glenn Drive
Amherst, New York 14228-2197

Published 2004 by Prometheus Books

Inquiries should be addressed to
Prometheus Books
59 John Glenn Drive
Amherst, New York 14228–2197
VOICE: 716–691–0133, ext. 207
FAX: 716–564–2711
WWW.PROMETHEUSBOOKS.COM

08 07 06 05 04 5 4 3 2 1

Library of Congress Cataloging-in-Publication Data

Waltar, Alan E. (Alan Edward), 1939–
Radiation and modern life : fulfilling Marie Curie's dream / Alan E. Waltar.
p. cm.
Includes bibliographical references and index.
ISBN 1–59102–250–9 (alk. paper)
1. Radiation. 2. Radioisotopes. 3. Radiation chemistry—Industrial applications. 4. Radioisotopes—Industrial applications. I. Title.

TP249.W35 2004
539.2—dc22

2004016023

Printed in the United States on acid-free paper

Contents

Acknowledgments

This book would never have been written without the original urging of Prof. Bernard Cohen of the University of Pittsburgh. He was the "fireplug" who recognized the need, based primarily on the pioneering work done by Roger Bezdek and colleagues about a decade ago that revealed the enormity of the previously unrecognized benefits of radiation technology to everyday life in America. Recognizing that this positive message needed to reach a wide audience, Bernie successfully convinced Dr. Glenn T. Seaborg, a Nobel Laureate who had personally served ten US presidents in high-level positions, to author the book. After all, Glenn was the codiscoverer of several of the radioisotopes now in common commerce—one of which is responsible for benefiting one out of every three Americans who enters a US hospital or medical clinic today. Hence, his name alone would certainly give such a book credibility and wide readership. Glenn agreed to do this, along with the help of knowledgeable colleagues in the variety of fields served by radiation technology. Unfortunately, Dr. Seaborg died before the project got formally underway.

Enter Linda Regan, editor of some of Dr. Cohen's previous books and currently executive editor at Prometheus Books. Fully appreciating the importance of the project, she pushed hard to find a way to resuscitate and complete the mission. Since I was one of those who had agreed to help Dr. Seaborg, Linda pressed me for several months on how to continue this undertaking and, in particular, how to attract support from someone who had the stature of Dr. Seaborg to provide the needed vision. One day I casually mentioned the visit of Marie Curie's granddaughter, Dr. Hélène Langevin-Joliot, to a special commemorative effort we held in honor of Marie Curie at Texas A&M University. Linda immediately recognized the possibilities, and she knew she had the makings of something very special. She convinced me to go to Paris and share these possibilities with Hélène. After careful consideration, Hélène agreed that the book was needed, and she graciously committed to both write an introduction and review the entire manuscript. That was the key to moving forward.

The only problem was that I had to agree to do most of the work! However, despite the long hours in trying to weave this project around a full-time job and numerous other commitments, I have found it to be an exceptional pleasure. Little did I realize how much our lives have been transformed for the better as a direct result of the pioneering discoveries of Marie Curie and those who followed her. Whereas I doubt that any living person has a full grasp of this impact, I have been the fortunate recipient of several very talented colleagues who agreed to review major portions of the manuscript in areas of their expertise to maximize both coverage and accuracy.

At the risk of missing key contributors, I would like to gratefully summarize the very constructive input and reviews by the following colleagues: opening chapter: Denis Beller; agriculture: James Dargie and Ray Durante; medicine: Bob Lull, Bob Schenter, and Darrell Fisher; electricity: Ann Bisconti and Angie Howard; industry: Peter Airey; transportation: Walt Laity and Roger Knight; space exploration: David Boyle and Fred Best; terrorism, crime, and public safety: Ned Wogman; arts and sciences: Dan Reece; environmental protection: Roy Gephart; modern economy: Roger Bezdek; and future glimpses: Lee Peddicord.

In addition to these special experts, I was pleased that Roger Bezdek, Bob Gehrke, and Don Todd agreed to review the entire manuscript. All three provided very valuable input. However, Bob Gehrke delivered true yeoman's duty. He was the only person I knew who had the breadth of experience to provide an in-depth review, and he did this with an incredible degree of professionalism.

Kathy Kachele deserves enormous credit for her masterful skill in creating the graphics. She has both the eye and the imagination of a true professional in transforming concepts into graphics that convey clear meaning. I am most grateful for her support.

I mentioned at the beginning that it was Linda Regan who really brought this project into being. I am most humbled by the confidence that she placed in me, and I'm even more grateful for the incredible patience and skill that she has demonstrated in transforming my bumbling drafts into coherent sentences and a logical thought pattern. She deserves much of the credit for making this book a reality.

But in the final analysis, it was the support of my nuclear family that gave me the freedom to spend the evenings, holidays, vacation days, and weekends necessary to complete this task. My dear wife, Anna, likely suffered the most from my focus on this effort, and it is to her that I gratefully dedicate the lasting impacts of this book.

Alan E. Waltar
Richland, Washington
February 2004

INTRODUCTION

Marie Curie and the World of Radiation
By Dr. Hélène Langevin-Joliot
granddaughter of Marie Curie and Pierre Curie
Director of Research Emeritus
National Center for Scientific Research
Paris, France

Radiation has existed since the very beginning of the universe. The story of the discoveries that led to our present understanding of the many different kinds of radiation started a few centuries ago. It focused first on the most important visible forms of radiation that constituted the sun light, and then extended to invisible radiation that we call ultraviolet and infrared radiation, the latter one being especially effective in heating homes and foods.

The scientific revolution that marked the end of the nineteenth century opened up a heretofore unknown world of radiation. It is striking that within only three years, several major discoveries dramatically changed the scope of science. From then on, experimental investigations were no longer limited to the macroscopic world. They extended to the microscopic world—not directly accessible to our senses.

For many years, the nature of the cathode rays produced by electric discharges in low-pressure gases had puzzled physicists. In 1895 Wilhelm Conrad Roentgen discovered that cathode rays produce a mysterious secondary radiation with extraordinary properties, x-rays. In 1897 Joseph John Thomson demonstrated that cathode rays were electrons—extraordinarily light particles charged with negative electricity. In the meantime, Henri Becquerel discovered a weak but spontaneous radiation emitted from uranium.

The discovery of x-rays immediately had a tremendous impact on the scientific community and on the public. The first radiography, taken by Roentgen of his wife's hand, was known all over the world within a few weeks, even days, despite word being sent through ordinary mail. The discovery of uranic rays did not appear at first as spectacular. Then, early in 1898, Marie Curie, my grandmother, discovered that the radiation is a property of the atom itself. That same year, she discovered with Pierre Curie two new elements, polonium and radium, which spontaneously emitted millions of times more radiation than uranium, and she coined the term *radioactivity*—the spontaneous emission of radiation. It was a turning point. From then on, Marie Curie's main preoccupation in studying radioactivity was to make radium and, more generally, the new elements and forms of radiation incomparable research tools, opening the way to dramatic breakthroughs in physics and chemistry, and later in medicine through radiotherapy.

The discovery of radium was indeed an outstanding scientific adventure. It was also a human adventure. The chance that the young Maria Sklodowska, born in Warsaw on November 7, 1867, would ever meet Pierre Curie, born in Paris eight years earlier, was small. But it happened. They were wed in July 1895 and started to collaborate just over two years later. Pierre was at that time a professor of physics at an engineering school for physics and chemistry that was operated by the city of Paris. In spite of this rather modest position, he was already recognized as a first-rank physicist. Marie had passed brilliantly her master's examinations in physics and in mathematics, and had just published the results of a first commissioned research work, yet she had no official position whatsoever. The discovery of radium in December 1898 became a hallmark in many respects. For the first time, a woman played a major role in science.

The young Pierre Curie had benefited from an education at home that gave special importance to science. His imagination and knowledge led him to become a scientist, while his sensitivity was more acute than that of an artist. He complained in his diary's notes: "Woman loves life for the living of it far more than we do" and also "Women with genius are rare."

The road toward scientific research was much more difficult for the young Maria, as a girl, and moreover as a Polish girl living in Warsaw under the domination of the Russian empire. Both her parents were professors. Her mother died when she was only eleven years old. Maria was a brilliant high school student, profiting from the encyclopedic knowledge of her father. The gold medal she received at the end of her secondary studies could not, however, open the doors of the university in Poland to her. She lacked money for studying abroad, as was also the case for her elder sister Bronia, who wished to study medicine. Giving private lessons in Warsaw was just enough to support themselves. They therefore decided to unite their forces to reach their goals.

Maria took a rather well-paid job for several years as a governess for a family with children, sending the main part of her earnings to Bronia until her sister would finally receive her medical degree in Paris. As the years passed, she nearly despaired until she could finally join Bronia. She was already twenty-four when she registered at the Sorbonne University of Paris, in the fall of 1891, to take courses in mathematics, physics, and chemistry. During the following three years she lived in an attic and worked days, evenings, and holidays to overcome her deficiencies. The result of her efforts surpassed her hopes. In a letter to her brother, she stated the rule that would govern all her life: "We must believe that we are gifted for something, and that thing at whatever cost, must be attained."

The small grant given to Marie for a study of steel magnetic properties prompted her meeting with Pierre Curie, in the beginning of 1894. As she discussed details of her project with him, he explained his recent results on magnetism. They appreciated each other at once. Both of them had ruled out love and marriage as part of their life's plans. But Pierre changed his mind very quickly, as they spent an increasing amount of time together. For Marie, however, her duty was to live with her family in Poland and to take a teaching position there. At the end of the university year, she left Paris. Pierre Curie wrote several letters, urging her to come back: "It would be a fine thing, in which I hardly dare believe to pass our lives near each other hypnotized by our dreams, your patriotic dream, our humanitarian dream and our scientific dream."

Fortunately for science, she came back. After a trip to the countryside in July 1895 after their wedding, they set up household in a small flat not far from the Sorbonne. Their first daughter, my mother Irene, was born in September 1897. Instead of searching for a teaching position in a secondary school, Marie decided to combine maternity, a husband, and scientific research.

Nearly two years before, Henri Becquerel had discovered the spontaneous radiation of uranium via the darkening of an exposed photographic plate. He had recognized that the rays were able to produce electric charges in air. The Curies had been puzzled by Becquerel's mysterious radiation. Marie decided to study this strange quirk of physics for her dissertation, adopting at once a quantitative method to perform her experiments—a change radically different from Becquerel's conceptual approach. As the codiscoverer of piezoelectricity, Pierre was ready to help set up an ionization chamber, coupled to an electrometer and a piezoelectric quartz, to precisely measure electric charges. After establishing without doubt the atomic origin of uranium radiation, Marie focused on her new program: searching to determine if the emission of radiation was, or was not, a general phenomenon by testing as many chemical elements and salts and even minerals as possible. The search turned negative for most elements, except thorium, but Marie noticed the very strange behavior of two uranium minerals in particular batches of pitchblende that emitted too much radiation. She dared to

Figure 1. Marie and Pierre Curie with daughter Irene in the garden of their home. (Courtesy of the ACJC-Curie and Joliot-Curie fund.)

state the far-reaching hypothesis that two unknown, very active, elements were responsible for the anomaly. From then on, Pierre Curie joined his efforts with Marie's to solve the enigma. The success came from their ingenious manner of combining chemical methods with quantitative measurements after each chemical separation to "trace" the unknown elements. This was the foundation of a new method of chemical analysis, the beginning of radiochemistry. She named the first radioactive element, discovered in July 1898, polonium (after her native country). The second one discovered in December 1898, just one year after the beginning of the work, was named radium (after rays).

It did not take long for the scientific community to recognize the huge new field now opened to researchers: the study of radioactivity. Within a few years, a number of discoveries by physicists and chemists of different countries, including Becquerel and the Curies, prompted the further exploration of this intriguing new world. It was observed, in particular, that the intensity of radioactivity changed with time, that additional radioelements existed, and that they emitted three different kinds of radiation. In Montreal, Ernest Rutherford and Frederick Soddy came to the astonishing conclusion that the emission of radiation resulted from the spontaneous transformation of one chemical element into another. The radiation released a huge amount of energy, as Pierre Curie had shown.

Discussions of a possible Nobel Prize for radioactivity and radium had begun to rumble early among Stockholm's Nobel committees. In 1903 the committee for physics chose a proposal that considered at first only Becquerel and Pierre Curie as recipients for the Nobel Prize, following the suggestion of the French Academie des Sciences. At the time of her discovery, Marie was indeed only a doctoral student, but above all, she was a woman! Fortunately, Pierre was privately informed of the committee's intention by a Swedish colleague. He protested that they both must be associated with the Nobel Prize, honoring the pioneering and original work of Marie. Accordingly, both Marie and Pierre shared the 1903 Nobel Prize for physics with Becquerel.

Anticipating the work that would lead her to another Nobel Prize in 1911, this time for chemistry, Marie had already begun focusing on the preparation of pure radium salts. Through sheer tenacity during years of laborious efforts, she succeeded in separating decigrams of radium salts, starting from tons of pitchblende residues. The only place she found to proceed, with the help of Pierre, to the most important steps of the separation was not in a laboratory, but in a simple shed, poorly heated and lacking hoods to evacuate radioactive vapors. In June 1903 in her thesis presentation, she announced the value of radium's atomic weight and designated its position in the table of chemical elements.

Marie later commented on her common life with Pierre. They enjoyed their family life with Pierre's father and with the young Irene, and later also with their second daughter, Eve, in addition to close friends. They spent days and many evenings at the laboratory, but they managed to stop working on weekends and holidays. The Curies believed that it was important to bring up their children in good health and for them to participate in sports. Tragically, the happiest moments of Marie's life were cut short on April 19, 1906, when Pierre was hit by a horse-drawn carriage on the streets of Paris and he died instantly. Her world was shattered without the love and companionship of her beloved husband. Fortunately, she had her two daughters to take care of. Pierre's colleagues at the Sorbonne, shocked by his death, recognized that Marie alone was the scientist who could succeed him. She resumed work as the first female professor and director of the laboratory devoted to research on radioactivity.

The separation of macroscopic quantities of radium had opened the way to its use as a source of powerful radiation. Marie Curie and Pierre had taken no patent for their radium extraction procedure. Their decision helped the development of a newborn industry that could produce increasing amounts of radium, make use of it in innovative applications, and especially answer the needs of radium for medical purposes. Preliminary results with radium rays and x-rays had already shown that radiation had usable biological effects. Marie Curie's research strategy during those years and until her death relied on her conviction

Figure 2. Marie and Pierre Curie in their chemistry shed. (Courtesy of the ACJC-Curie and Joliot-Curie fund.)

that strong radiation sources were the key to new discoveries as well as to important progress in cancer treatment. Her fight for an institute in Paris, dedicated to this research with a section for physics and chemistry and a section for biology and medicine, was finally successful, but the laboratory was not completely finished when the First World War broke out in 1914.

Marie Curie's main preoccupation was then not radioactivity, but the organization of radiology and radiotherapy services for military hospitals—in particular, setting up mobile radiological posts on cars or vans. She focused her attention on the needs in the army zones, and upon examining the wounded soldiers, explaining how to best use x-ray apparatuses. Often at her side, either

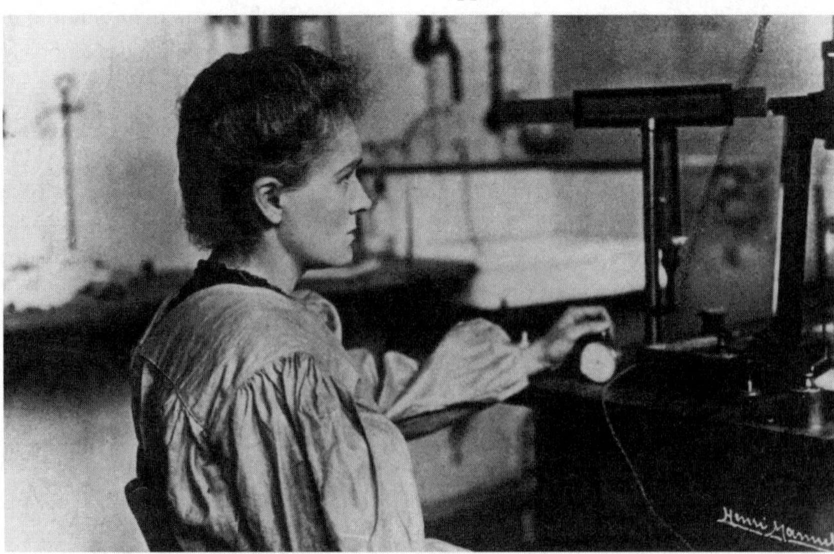

Figure 3. Marie performing a radioactivity measurement. (Courtesy of the ACJC-Curie and Joliot-Curie fund.)

on the battlefront or at the Radium Institute where Marie had organized efficient training sessions for nurses, was my mother, Irene.

When the war was over, the Radium Institute suffered from the lack of money in France. It did not have enough radium for even research at the very moment powerful alpha particle sources were needed in the laboratory to play the role that modern accelerator beams not available at that time would later play. Moreover, radium gamma rays were developing as an efficient method to fight cancer, and larger amounts of radium were needed. An unexpected source of help came in 1921, with the visit of Mrs. William Brown Meloney. An editor of a women's magazine in the United States, she was shocked by the situation and proposed to organize a subscription campaign. Marie traveled to the States. The round-trip culminated with her reception at the White House, where President Warren Harding presented her with one gram of radium, on behalf of the women of the United States. She came back with additional funds, equipment, and radioactive products. Mrs Meloney and Marie stayed close friends for the remainder of their lives.

The Radium Institute became a modern laboratory, comparable to any among the best in the world. Marie's experience, in collaboration with physicians, proved especially instructive in the creation of the Curie Foundation. The foundation became a world-renowned leader in radiotherapy.

Marie Curie managed the institute, gave advice about works in progress, and

Figure 4. Marie training American soldiers on the practical uses of radiation at the Radium Institute. (Courtesy of the ACJC-Curie and Joliot-Curie fund.)

Figure 5. Marie in the United States (in front of a factory in Pittsburgh). (Courtesy of the ACJC-Curie and Joliot-Curie fund.)

pursued her own research. My mother, Irene, became her assistant, and she married a young scientist, Frederic Joliot, who worked as a researcher for Marie. In January 1934 Frederic and Irene Joliot-Curie announced their discovery of artificial radioactivity. Using the alpha particle of a strong polonium source, they had transformed aluminum into a radioactive isotope of phosphorus. It thus happened that a second generation received a Nobel Prize, for chemistry in 1935. Artificial radioisotopes opened the way to the atomic age. Because of their special radiation properties, radioisotopes had remarkable applications in biology and medicine. When Marie Curie died in July 1934, her scientific dream and her humanitarian dream had merged together to become a reality.

But could Marie have ever imagined how her scientific achievements would lead to such huge direct and indirect consequences at the eve of this new century? What would she really think if she were alive today? We will never know. The only thing I can say with reasonable confidence is that she would be sad knowing that many people still fear radiation today, and that they are not aware of its enormous benefits.

It is true that we cannot forget the mushroom clouds above Hiroshima and Nagasaki. The incredible power of nuclear arms, if recklessly used, could destroy life on Earth. We also cannot forget the tragic Chernobyl accident. This has led many people to fear nuclear power plants, and some even think there is no way to properly handle radioactive waste.

But the answer is not to fear. Rather, it is to understand. Many technologies have beneficial attributes that can become dangerous if improperly used or not controlled. We need to continue with the scientific research necessary to achieve solutions that will optimally benefit society.

Figure 6. Marie and Prof. Claudius Regault, celebrating the fight against cancer at the new Radium Institute in Warsaw, Poland. (Courtesy of the ACJC-Curie and Joliot-Curie fund.)

Marie Curie liked to emphasize that science has a great beauty, that scientific breakthroughs were driven by the spirit of adventure and curiosity. She also contemplated the possible misuses of science, as stated in Pierre Curie's Nobel lecture:

> One can imagine that in criminal hands, radium could become very dangerous, and here one must ask oneself if humanity gains anything by learning the secret of nature, if humanity is ready to profit from this or whether such knowledge may not be destructive for it. I am one of those who thinks like Alfred Nobel, that humanity will draw more good than evil from new discoveries.

Let us use Marie Curie's discoveries for the greatest benefit to humanity.

Figure 7. Marie with daughter Irene and granddaughter Hélène. (Courtesy of the ACJC-Curie and Joliot-Curie fund.)

Figure 8. Marie Curie, 1867–1934, "Nothing in life is to be feared. It is to be understood." (Courtesy of the ACJC-Curie and Joliot-Curie fund.)

PROLOGUE

The discovery of radiation just over a hundred years ago by Marie and Pierre Curie was undoubtedly one of the most remarkable events of all time. It has transformed the worlds of mathematics, physics, and chemistry and, as such, has completely reshaped our modern environment. By learning how to exploit radiation, we have been able to make remarkable strides toward improving the quality of life throughout the world.

By a great stroke of luck, I recently had the privilege of meeting Dr. Hélène Langevin-Joliot, the daughter of Irene and Frederick Joliot and the granddaughter of Marie and Pierre Curie. Hélène is not only a gifted high-energy physicist in her own right. She is a most charming and engaging person. She agreed to participate in a special project that I helped organize at Texas A&M University in 2000 entitled "Women in Discovery: Celebrating the Legacy of Marie Curie." This event captivated the large institution, drawing direct and enthusiastic support from nearly all the colleges on campus.

One of the key features of the project was the creation of a museum-quality exhibition to be housed in the J. Wayne Stark Gallery, located within the Memorial Student Center, in the heart of student activity. With the direct support of Dr. Langevin-Joliot, we were able to display some of the original artifacts that Marie Curie used in her research that led to her two Nobel Prizes. It was the first time this archival equipment had ever been on US soil. As a result of the tremendous reception that Hélène and the exhibition received, she graciously allowed Texas A&M curator Catherine Hastedt to send the exhibit to ten sites within the United States over the next two years. The exhibition drew over 230,000 visitors as it crisscrossed the nation.

Figure 9. Original instruments used by Marie Curie for her Nobel Prize work (on display at the Women in Discovery celebration, Texas A&M University, March 2000).

As an interesting sidelight, all of us were holding our breath in anticipation of what Hélène would think of our presentations in the exhibition. She arrived the evening before the grand opening and was given a special preview of the layout. Her first reaction was one of shock when she noted that we had mislabeled the original artifacts from her famous grandmother's laboratory. Needless to say, we made the necessary corrections in very short order!

Interestingly, she was not nearly as impressed with the artifacts as we were. After all, she grew up with them—spending countless hours in the laboratories of both her parents and that of Marie Curie herself. What really captivated her interest was a small vial of radium that had been hand-packed by her grandmother. Only weeks before the grand opening, we learned that a vial of radium had been discovered by the Department of Energy (DOE) at a home in Grand Junction, Colorado, once owned by Walter Koenig. As the story unfolded, we learned that Mr. Koenig had worked in Marie Curie's laboratory in the early 1900s and later had returned to the United States to start up a uranium mining company in Colorado. He was very instrumental in helping Marie raise money during her long visit to America in her quest for radium. She was so appreciative of his help that she gave him this twenty-five-milligram vial of radium-226 (a vial very similar to ones she hand-packed for medicinal purposes). The DOE, having learned of the special "Hundredth Anniversary of Radiation" celebration that Texas A&M was hosting, offered the vial to the Department of Nuclear Engineering of our school. Needless to say, we jumped at the opportunity. As Hélène gazed at the vial, protected with thick leaded-glass shielding, she remarked that as far as she knew, this was the only original vial still in existence. A most interesting coincidence.

Starting with this special relationship with Dr. Langevin-Joliot, I have been privileged to enjoy several subsequent conversations with her, both in the United States and in Paris. Though we come from very different backgrounds and cultures, we found that we both share a common passion: namely, to help a wider segment of the public better understand and appreciate the marvelous benefits that ionizing radiation has brought to modern life.

Figure 10. Hélène Langevin-Joliot with students at the Women in Discovery celebration, March 2000.

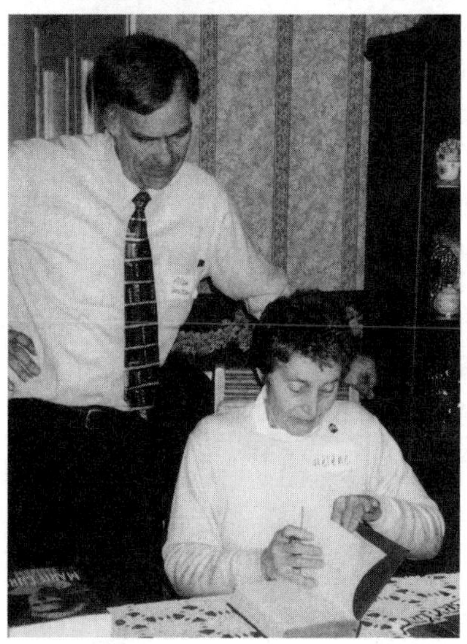

Figure 11. The author with Dr. Langevin-Joliot at the Women in Discovery celebration, March 2000.

Many people today fear the word *radiation*, largely because of images of a mushroom cloud or the pictures of horror that filled global newspapers following the tragic accident at Chernobyl. In the aftermath of 9/11, many citizens fear that harmful radiation will be dispersed by terrorists who are threatening to explode a "dirty bomb." Such imagery is frightening, indeed.

But we must remember that there are many things in modern society that can be dangerous if used in excess or if uncontrolled—and yet we have found ways to harness them for enormous good. Fire is but one example. Fire has destroyed hundreds of thousands of acres of prime timberland within the past few years. It has likewise ravished major portions of cities and claimed thousands of lives over just our own lifetimes. Yet fire, properly controlled, is an absolutely essential part of our everyday existence. Without fire we would not have internal combustion engines (to power our cars and trucks) or jet engines (to fly our airplanes), nor would many of us be able to heat our homes and workplaces, or even be able to cook our meals.

Viewing radiation from this perspective, we soon grasp the necessity of looking at it in its fullest and fairest sense. Radiation is sufficiently powerful to provide enormous good or enormous harm. It is not intrinsically either good or bad. It is how we choose to use it. We need to acknowledge straight away that Marie Curie died at age sixty-six (on July 4, 1934) from a lifelong exposure to high doses of radiation. The combination of her almost daily exposure to relatively intense sources of radiation, an inadequate knowledge of the health impacts of high doses at that time, and the lack of modern shielding techniques were prime factors in cutting her brilliant life short.

Yet as Marie Curie once said, "Nothing is to be feared. It is to be understood." Once we ponder the implications of the pages to follow, I hope we will all better appreciate how much we are indebted to the price that she paid for her devotion to laying the cornerstones for understanding what has become one of our most valuable modern tools. It is within this spirit that I encourage you to open your mind and join me in a tour of how this age-old phenomenon has been harnessed over the last century to bring almost-unfathomable advances and richness to our modern way of life.

1. Our Common Quest for Life

Chicago, Illinois *

Stacey Toscas was frustrated. With a family of six, including two teenage boys who seemed to consume the equivalent of at least one bale of alfalfa per day, it seemed impossible to keep a stash of fresh fruits and vegetables in her kitchen. Stacey normally purchased such items in bulk quantities to satisfy the seemingly bottomless pits that surrounded the family dinner table. Still, there were inevitably times when one or more didn't show up for dinner, given all the extracurricular activities surrounding twenty-first-century living—football practices, special orchestra rehearsals, concerts, and tennis lessons.

As a consequence, large quantities of strawberries, raspberries, onions, potatoes (you name it) had to be discarded because of mold, rotting, or sprouting. In addition to fielding gripes caused by having no fresh strawberries for the morning bowl of granola, Stacey was frustrated by the real expense of tossing out so much inedible food. She didn't like having to dispose of large, wasted quantities of produce. A teacher by profession, Stacey often contemplated the ethics of our throwaway society. She was well aware of the extent of starvation that plagues many parts of the globe, so she felt guilty every time that she was forced to discard sizable quantities of spoiled food.

Hence, Stacey was intrigued when she learned that the Carrot Top grocery store in northern Chicago was selling a range of fruits and vegetables that were

* The first three narratives in this opening chapter are composites of stories of several real people. The last three narratives are completely true life stories.

advertised to have much longer shelf lives than regular produce. Could this be for real? Could strawberries really last for weeks, rather than days, without refrigeration? Could potatoes be stored in the corner of the garage throughout the entire winter without developing sprouts?

Though a little skeptical, she could hardly contain her enthusiasm when she read the literature prominently displayed in the special produce corner that contained this ultrafresh array. The "special" products all looked great, especially in comparison with the produce in the adjacent bins—fruits and vegetables that had not been given the new treatment. Stacey decided to conduct her own experiment by purchasing a few cartons of these advertised strawberries and blueberries, along with a small sack of "preserved" onions.

To her great delight, all of the claims turned out to be true. Having purchased similar quantities of "natural" or conventional produce, she had a side-by-side comparison. After feeding "unprocessed" and "processed" produce to her family for several days, she was convinced that the flavor and overall enjoyment between the two was indistinguishable. The only major difference was that the "processed" foodstuffs lasted for weeks (without refrigeration!), whereas the conventional fruits and veggies began molding within days as usual—even if they were kept chilled. That was it. Stacey quickly converted her entire produce budget to these impressive "processed" foods.

Now what was this special "processing?"

RICHLAND, WASHINGTON

Roy Gephart anxiously awaited the first NASA pictures from the *Voyager 2* as the spacecraft activated its camera and began focusing on the spectacular rings of Saturn. Four years earlier, *Voyager 2* had blasted off from the Kennedy Space Center aboard Titan-Centaur rockets and began its interplanetary trek. It was followed by its sister spacecraft, *Voyager 1*, just two weeks later.

The whole process was of special interest to Roy. He had been fascinated with the heavens since his boyhood days of growing up in Miamisburg, Ohio. Though his eyes were often lifted upward to the starlit sky, his formal education was firmly tied to Mother Earth; namely, the geosciences. Having honed a specialty in groundwater hydrology, he was drawn to the Hanford area in eastern Washington to help the US Department of Energy explore ways to return vast portions of the site to its original environmental condition. That site was contaminated in several areas due to a half century of use for military weapons production.

An additional lure of the Hanford region was the vast, open sky of the desert, which allowed Roy to cultivate fully his real passion—astronomy. As such, he soon became the astronomy "sage" of the community, organizing the local astronomy club, holding stargazing sessions for the public and school children, and often writing guest newspaper articles to promote stargazing among members of the community.

So, it is no wonder that Roy took such interest in the Voyager project. He had studied for some time the rings of Saturn through his Maksutov home telescope—always longing for a closer image. His tastes were further whetted while spending countless nights peering through the thirty-one-inch Cassegrain telescope featured at the nearby Rattlesnake Mountain Observatory.

But this special day was his first opportunity for a real close-up of the Saturn rings. The initial pictures were a bit fuzzy, perhaps not unexpected considering that the images were being returned some eight hundred million miles back to Earth. But as the days flew by, the crispness improved until the images created an absolutely breathtaking sight.

As an engineer, Roy realized that substantial power was required to capture these images and then to send them back over such a vast void to Earth. Was this solar energy? Certainly not. On this mid-Autumn day of 1981, during its closest approach to Saturn, *Voyager 2* was over nine hundred million miles from the sun—where the impinging solar energy would have been roughly 1 percent of that striking Earth. The amount of solid or liquid fuel that would have had to be transported to power such a mission would be extremely expensive. Further, *Voyager 2* had already been "on the road" for over four years, with a flyby past Jupiter, and it was still destined for Uranus and Neptune before heading into interstellar space—a multidecade mission. Clearly, the time demand for capturing and transmitting such pictures would have ruled out the role of conventional batteries.

What in the world could provide the power demands for such a project?

Quito, Ecuador

Roger and Carol Johnson could hardly wait to get to Ecuador where their son, Scott, was serving in the Peace Corps. Scott was their only child, and they were justifiably proud of his concern for the less fortunate in the world. Hence, they had not hesitated in giving their approval for his desire to help improve living conditions in this impoverished South American country.

After six months of being apart, they decided that they wanted to pay him a visit. Elaborate planning was necessary to acquire and package the special items they knew he needed for his work. Finally, they met all the deadlines and soon after collapsed in their seats just in time for takeoff.

They knew the flight schedule would be grueling, given the 8:00 AM departure from Seattle, with stopovers in Los Angeles and San Jose, Costa Rica. The last leg was particularly beautiful, however, with the panorama of the majestic Atlantic Ocean on the left and the glimmering Pacific Ocean on the right as the sun began to set.

After a quick nap, they were awakened by the routine stewardess check to push their seats forward and prepare for the final approach. They could see the lights of Quito in the distance, piercing the night sky like a sea of crown jewels to welcome their arrival. The landing gear was lowered and their pulse quickened, knowing that in only about twenty minutes they would be reunited with their son after more than six months of separation.

Then, without warning, it happened. All at once the thousands of twinkling lights below them went blank. The earth below was a solid abyss of blackness. They rubbed their eyes, wondering if they had just encountered a thick cloud cover (though all weather predictions were for a cloudless sky). Other passengers began to murmur and press their noses to the windows with equal astonishment. After about five minutes of anxiety, they felt the roar of the rethrottled engines, and the airplane lunged upward. They slowly banked to the left and seemed to be settling into a circular pattern.

Finally, the captain activated the public-address system and acknowledged to the passengers that there was a bit of uncertainly about what was going on. All contact to the tower had been lost—a situation that he had never previously experienced. He explained that all modern airports had backup systems for loss of power, but given the present cutoff, plus the sudden loss of lights from the city, something exceptionally unusual was underway. He hesitated to speculate, but he did suggest the possibility of a massive earthquake—perhaps of such magnitude that it would have ruptured even the electrical service from the backup diesel generators. Indeed, Ecuador was known as earthquake country.

So what could the pilot do? The captain assured the passengers that they still had enough fuel to make alternate landing arrangements. While reassuring to most passengers, Roger had done his homework and knew there were not many airports in that part of the world with runways sufficient to land an airplane of this size. Furthermore, at the altitude of Quito, it was necessary to fly at a relatively high airspeed to maintain control in the thin atmosphere. Hence, trying for an emergency landing at a small airport would be exceptionally risky.

After another ten minutes of circling, the captain came back on the air to say that a massive earthquake had been confirmed. However, he announced that he had decided to attempt a landing at Quito—even without contact with the tower. The visibility was good *and* the main runway had working lights along the full landing strip!

Whereas there were many white knuckles in the passenger compartment, the captain landed the plane without incident. In fact, it was one of the smoothest landings that Roger and Carol had ever experienced. The subsequent hugs with Scott upon their arrival were doubly meaningful. He, too, was safe.

With all power lost to the city, the airport, and the control tower, how could those lights still be burning along the runway? What kind of technology did this airport have that could have survived a massive earthquake and allowed that airplane to land safely?

JERUSALEM, ISRAEL

It is an old adage that one of the best ways to get into an argument during a family gathering is to discuss either politics or religion. Perhaps the more passionate of the two is religion. And so it is no wonder that attempts to prove the historical accuracy of religious writings is fraught with controversy.

One of the most interesting arguments over the past several centuries is the authenticity of the biblical claim that King Hezekiah built the famed water tunnel that provided water to the city of Jerusalem in anticipation of a siege by an invading Assyrian army. If the biblical account is accurate, the tunnel would have had to have been built sometime during his reign from 715 to 686 BCE.

Some scholars have argued that the tunnel was built about five hundred years later. And there is a legitimate basis for arguing that such a feat could not have occurred as early as the biblical accounts state. Indeed, this tunnel is nearly 1,750 feet long and tall enough for a man to walk fully erect inside. It twists and turns in its path from the Gihon spring, located outside the city walls, into the Siloam pond situated inside the walls. As such, this tunnel is generally considered one of the greatest engineering feats of all time.

In an attempt to settle the controversy, Amos Frumkin of the geography department at Hebrew University of Jerusalem led a team of scientists to this popular site (still a major tourist attraction) and removed some of the original plaster from the walls and floor of the tunnel. He was amazed to find that some well-preserved plant fragments must have been accidentally incorporated into the plaster as it was mixed outside the tunnel. He then took some of these small, preserved plant fragments to a special laboratory for testing. When the results were returned, he triumphantly declared that the tunnel was indeed built in 700 BCE, plus or minus a hundred years.

His measurements, corroborated by further measurements of the stalactites formed in the ceiling of the tunnel where water had seeped in through cracks and fissures, have conclusively ended this centuries-old controversy. Old King Hezekiah was quite a builder!

What in the world was the measurement technique used to provide such a degree of authenticity?

DAYTONA BEACH, FLORIDA

There is no question about it. Paul Newman loves acting and making movies. His lifetime achievements—ten Academy Award nominations—provide ample testimony to his success in his chosen profession. He received an Honorary Oscar in 1985 in recognition of his many compelling screen performances, for his personal integrity, and for his dedication to his field. The next year he won the Oscar outright for his performance in *The Color of Money*, and he's been nominated twice since then as well as receiving a Humanitarian Award for his philanthropic work.

But even more exciting to Paul than performing in front of lights and cameras is that of racing around an asphalt track at 180 mph in his Sports Car Club of America (SCCA) Jaguar, competing against drivers one-fourth his age, and winning. That's appropriate, because *Winning* was the title of the movie that got Paul's racing career started. For more than three decades, long weekends have often found Newman, with or without Newman-Haas Racing—his world-champion Champ Car racing team—working with Ford Motor Company in a search for higher-performance racing engines and actually testing them on racetracks all over the world.

Paul's first interest in car racing came later in life than most racecar drivers. When he was forty-three years old, he had the lead role in the 1968 movie titled *Winning*, playing a championship Indy Car driver who was fighting for the title and to keep his wife. His real wife for more than four decades, his lifelong partner and true love, the Academy Award–winning actress Joanne Woodward, also played his wife in the film. During the filming, Newman fell in love again, this time with auto racing. The screaming tires, roaring engines, exhaust with unburned methanol fuel, and wheel-to-wheel racing at speeds over two hundred miles per hour were a little too much to resist. Wow! What an atmosphere! What power! What speed! He immediately became truly fascinated, and he couldn't get started in real competition soon enough. But his filmmaking career, making three to four pictures a year at that time, made it tough to get his racing career going.

Newman's racing career began three years later, in 1972, driving a Lotus

Elan in a SCCA event. His first national SCCA championship came in 1976, and he followed that with national championships in 1979, 1985, and 1986. The pinnacle of Paul's racing career was probably his victory as a codriver in the GTS class of the Rolex 24 Hours of Daytona on February 5, 1995, driving a Ford Mustang. It was a great way to celebrate his seventieth birthday. The several Indy Car and Champ Car World Championships that his team's drivers have won, which includes victories by Mario and Michael Andretti, and Nigel Mansell, as well as Bruno Junqeira, probably haven't dampened his enthusiasm, either. Paul Newman is still driving racecars at the age of seventy-eight, and he's still beating those youngsters.

Years after his introduction to racing, Paul would learn that the engines of his youth were highly inefficient, and as any race fan can verify, they surely didn't last very long. The blowby around the pistons not only polluted the atmosphere; it substantially reduced the usable power transmitted to the wheels. Through a special new technology, it became possible to increase the efficiency of these devices substantially and to build engines with a power-to-weight ratio unheard of before. In addition, the wear of parts could be monitored without removing them from the engine. Given this technology, Paul Newman, his Champ Car crews and drivers, and his colleagues have been mesmerizing millions of fans worldwide with performance and endurance previously thought to be impossible.

What is this marvelous new technology?

CHEHALIS, WASHINGTON

It was one of those calls that we hope to never get. My mother was calling to say that she had just been diagnosed with cancer. Bone cancer! My stomach involuntarily flexed inward as if it had just been hit with a bowling ball—and my throat went dry. Surely this could not be.

My mother had always been so active. She was typically up at 5 AM to till and hand-water her garden, mow the lawn, can peaches, and then bake a half dozen blackberry pies for the Grange dinner that evening. Following Dad's stroke at the early age of fifty-one, she became the principal family breadwinner, laboring long hours at the county auditor's office as the chief bookkeeper. After Dad died, it seemed that she increased her pace—working circles around everyone in her immediate environment, yet doing this all with a selfless heart. Though never one to seek praise, she was widely recognized as one of the most compassionate, energetic women in the community.

Was the life of this human dynamo about to be cut short with cancer? And of all things, bone cancer—one of the most painful of all forms of this hideous disease?

After pausing to get a grip on my emotions, I mumbled something to the effect that we would seek a second opinion. After all, my mom was still living in the old two-story farmhouse near the small rural town of Chehalis, Washington. Although the local doctors were competent, I quickly tried to convince myself that she might benefit from going to Seattle, where doctors had considerably greater degrees of specialty.

We arranged for her to see an oncologist in Seattle. Only minutes after he injected her with a special medication, the doctor brought the family together and gave us the news. "Mrs. Waltar," he exclaimed, "you are a fortunate woman." He pointed to a picture that looked strangely like an x-ray, but with dashing splashes of color. "You do not have the hot spots characteristic of bone cancer. I suspect it is simply inflamed arthritis."

With a great sigh of relief, we drove Mom back to her home and celebrated with a hearty meal of country-fried chicken, topped off with wild blackberry pie a la mode. Sure enough, she lived many productive years after that scare, with never a sign of cancer. How in the world could that Seattle doctor have been so confident in reversing the near-paralyzing diagnosis of Mom's general practitioner from Chehalis?

What was that magic injection?

Fresh-food preservation, spacecraft power, airport lights, historical dating, efficient engines, cancer detection. What can they possibly have in common?

The answer. Radiation.

Radiation? Yes, ionizing radiation. How could this possibly be? Isn't radiation that stuff we're so worried about in nuclear waste? Radiation is indeed an integral part of nuclear waste. But radiation, and ionizing radiation in particular, is a phenomenon that has been around for billions of years. It is only within the last century that humans have been able to harness it for a plethora of useful and many lifesaving applications. As we shall soon see, it is practically impossible to go through a day without directly *benefiting* from an enormous array of products or processes made possible by exploiting the awesome properties of radiation. In fact, harnessed radiation in the United States today adds more than an astounding $400 billion to our national economy annually, as well as over four million jobs.

I invite you to take a ride with me to explore some of the ways that this recently discovered phenomenon has so quietly but effectively enriched our lives.

2. Thriving in Radiation

The word *radiation* is perhaps one of the most misunderstood words in the English language. To scientists and engineers, radiation is now well recognized as a natural phenomenon that has existed even longer than the age of our Earth. But to many people, radiation is often thought to represent a new twentieth-century "force" introduced only recently by the nuclear age. My intent is to help take the mystery out of radiation. An additional purpose is to explain how this age-old phenomenon has been exploited over the last century to become one of the most powerful tools for enriching our daily lives.

Radiation is as much a part of our natural surroundings as air and water. In fact, it is far more prevalent than air and water since there is literally no place on Earth, or in the universe, where radiation does not exist.

What Is Radiation?

Radiation is, in its most generic sense, simply energy traveling through space. Sunshine is probably the most familiar form of radiation. Sunshine is composed of photons (packets of energy) that emanate from the sun in the form of rays. These rays are essential for sustaining plant life, for vaporizing water to develop our clouds (thereby forming the rain cycle), and for heating and lighting Earth. Radiation is essential for life as we know it.

But like everything else, too much sunshine also carries a risk. Most of us have at one time or another soaked up too much of these warm summer rays and gotten a sunburn to show for it. The rays of sunshine come in a range of energy packets, called photons. The energy of a particular photon is inversely proportional to the length of its wave. In other words, the lower the energy in the photon, the longer the wavelength. We comfortably see the longer wavelengths with our naked eye, but it is best to wear sunglasses with special lenses to shield our eyes from the shorter-wavelength (higher-energy) ultraviolet rays. In fact, it is the more energetic ultraviolet rays that cause the sunburns—rays that we can't even see.

Figure 12 illustrates the range of wave frequencies covering what is called the energy spectrum. Low-frequency radiation, characterized by the lowest-energy (longest-wavelength) rays, is the common radio wave. The next higher-energy rays are called microwaves, which have now become familiar to most households via our kitchen microwave ovens. Next come the infrared rays, sometimes used for space heating or special cameras designed for night viewing. We now arrive at the visible light spectrum that emanates from the sun or a lightbulb. That is the light we can see with our own eyes without the aid of any special instrumentation. The full palate of brilliance is on display when we shine visible light through a prism or, better yet, grasp its full majesty in the appearance of a rainbow. However, as noted above, it is the highest-energy component (the short ultraviolet rays) that can cause damage to our eyes or skin.

Figure 12. The energy spectrum.

None of the energy ranges described above have sufficient energy to cause chemical changes in the materials they strike, although they can certainly heat materials—even to the point of causing burning, where chemical changes do take place. But at higher energy, radiation can become manifest both as rays in the form of photons and as particles, such as electrons. At these high energies, there is sufficient energy to knock an electron out of its atomic orbit, causing the struck atom to become an ion.* This is called ionizing radiation, a process that can lead

* An ion is an atom that lacks one or more electrons in its outer orbital shell. Because of this loss of electrons, an ion carries an electrical charge.

to substantial chemical changes. High-frequency radiation, including x-rays, gamma rays, and cosmic rays, is in the category of ionizing radiation. Many of these photons and particles are spontaneously emitted by radioactive substances that relate directly to the behavior of the atomic nucleus, and they are capable of doing damage to the human body. However, their power can also be harnessed for a wide variety of beneficial uses. That is why we shall focus on this type of radiation.

Accordingly, we probably need to back up a bit and make sure we are all aboard on understanding the basics of atomic structure.

All *atoms,* the smallest entities in nature that have unique chemical and nuclear properties, are comprised of a nucleus at their center and orbiting electrons. Embedded in the nucleus are protons and neutrons. These particles are of almost-identical weight, but the protons carry a positive electrical charge whereas the neutrons have no electrical charge at all. Since protons and neutrons are nearly two thousand times heavier than the negatively charged electrons that orbit the nucleus, essentially all of the weight of the atom resides in the compact nucleus.

Hydrogen, the simplest atom, normally consists of a single proton in the nucleus (no neutrons) and one orbiting electron. Since the proton has a positive electrical charge and the orbiting electron has an equally negative electrical charge, such an atom is electrically neutral. The two opposite charges simply balance each other. The atom we have just described is the *element* hydrogen. It is important to note at this point that the chemical properties of an atom are determined by the number of protons in the nucleus, whereas the nuclear properties are determined by the combination of protons and neutrons in the nucleus.

If, for some reason, the orbiting electron should be removed from the hydrogen atom described above, this atom would become known as a hydrogen ion, and this ion would have a positive charge. One way for this to happen would be for some type of ray or particle to interact with this electron with sufficient energy to literally remove it from its orbit, sending it well away from the positive attraction offered by the proton in the hydrogen nucleus. Such an event, removing an electron from its orbit, is called *ionizing radiation.*

Let's continue to look at atomic structure and consider some of its nuclear properties. We mentioned that hydrogen normally has only a single proton in its nucleus. But it sometimes appears with a neutron in its nucleus—in addition to its signature single proton. Since essentially all of the weight of the atom is comprised of the protons and neutrons in the nucleus (and protons and neutrons have about the same weight), the atom we just described has twice the weight of the normal hydrogen atom. However, since the chemical properties of an atom are determined solely by the number of protons, this new species is chemically identical to the normal hydrogen atom. We call this an *isotope* of hydrogen. This isotope of hydrogen has been given the name deuterium. There are hydrogen atoms found in nature (and they can also be manufactured) that even have two neutrons within their nucleus, in addition to the signature single proton. These atoms are called tritium. Again, all three of these species are essentially brothers (or sisters). They are isotopes of hydrogen. They are all the same element, since they have identical chemical properties (because of their single proton). But they have vastly different nuclear properties—because of the fundamental differences in their nucleus.

If we look at the periodic table (the orderly listing of elements that make up all of nature—which we have all seen in a high school science room), we note that helium, the next atom that we encounter after hydrogen, normally contains two protons and two neutrons in its nucleus, and its nucleus is surrounded by two orbiting electrons. Again, all atoms with two protons in their nucleus are the element helium, but there are instances where the number of neutrons in the helium nucleus differs from two. These are examples of different isotopes of helium (i.e., same number of protons but different numbers of neutrons).

By examining the entire periodic table, we soon discover that all of the elements in the universe include two or more isotopes. Many of these occur naturally, but many more are man-made. Perhaps of greater interest, over half of the elements in our universe contain at least one naturally occurring isotope that is unstable. An unstable atom is one that will, at some random time in the future, spontaneously change to a different element. When the nucleus of an atom spontaneously changes, it gives off energy in the form of radiation. Thus, isotopes that change to an isotope of a different element are called *radioactive*, and the process of spontaneous change is called *radioactive decay*. Isotopes that are radioactive are often referred to as *radioisotopes*. This process occurs when the unstable nucleus ejects particles as it seeks to achieve a lower energy state (more stable) nucleus.

Sometimes the unstable nucleus ejects an *alpha particle*, which is simply a helium nucleus consisting of two protons and two neutrons. This changes the decaying radioisotope into a new element (one with two fewer protons and two fewer neutrons).

In other situations, the unstable isotope may release an electron (called a *beta ray* or *beta particle*)* from the nucleus. This also changes the radioisotope into a new element (a species with an additional proton and one fewer neutron in

* Beta emissions can be characterized either as particles or as electromagnetic energy rays, depending upon the mode of interactions being studied.

its nucleus). The beta particle ejected from the nucleus usually has a negative charge because it is an ordinary electron, resulting from the transformation of a neutron into a proton. In a few instances, however, the beta particle carries a positive charge. It has the same weight as an ordinary electron, but it results when a proton is transformed into a neutron. This particle is called a *positron*. In this case, the resulting element has one fewer proton and one additional neutron.

In yet other situations, the radioactive decay process results in the release of a photon (a *gamma ray*) from the nucleus. In this case, both the chemical and the nuclear properties of the resulting isotope remain unchanged. However, gamma rays are often emitted during alpha or beta decay processes as well, as a mechanism to satisfy energy balances.

On rare occasions, neutrons are emitted as a radioactive decay mechanism. Californium-252 is an example of a radioisotope that decays in this mode as a result of spontaneous fission. Our principal interest in neutrons, however, relates to the fission process that occurs in a nuclear power plant, where several neutrons are released along with other nuclear particles.

When radioactive decay occurs, the time required for 50 percent of the unstable isotopes (radioisotopes) to disintegrate (i.e., lose one-half of the original level of radioactivity) is known as a *half-life*. Stated differently, after one half-life, only 50 percent of the original material is still in existence, and after two half-lives, only half of the half (or 25 percent) of the original element is still around. It is interesting to note that the half-lives of various substances differ widely, from very small fractions of a second to billions of years.

Some people become concerned when they learn of a radioactive substance with a very long half-life, say a million years, because it sounds like it will be dangerously radioactive forever. In reality, such a substance is likely to be much safer than one with a moderately short half-life, because atoms with shorter half-lives are decaying much faster, thus emitting radiation energy at a considerably faster rate.

Even though this radioactive process has been occurring since the beginning of time, ionizing radiation, as we've noted, was not discovered until about a hundred years ago. The reason this phenomenon was able to escape observation for so long is because ionizing radiation cannot be detected by our senses (sight, sound, touch, taste, smell). Rather, special detection equipment is needed. One of the great accomplishments of the Curie team was to develop instruments that had the capability to measure this previously unknown phenomenon.

In short, we now know that there are four principal types of "particles" emitted during a radioactive decay process: alpha particles, beta particles (plus positrons), gamma rays, and neutrons.

Alpha particles, which we noted above to contain two protons and two neutrons, are identical to a helium nucleus (a helium atom would also contain two orbiting electrons), except that alpha particles are normally moving with a large kinetic energy. Since alpha particles are the heaviest of these four particles, one would intuitively suspect that they would have the most penetrating power through matter. Ironically, however, they are the easiest to shield against because they contain a highly positive electrical charge and are readily stopped. Even a thin piece of paper or an outer layer of skin provides an effective alpha particle shield (see figure 13).

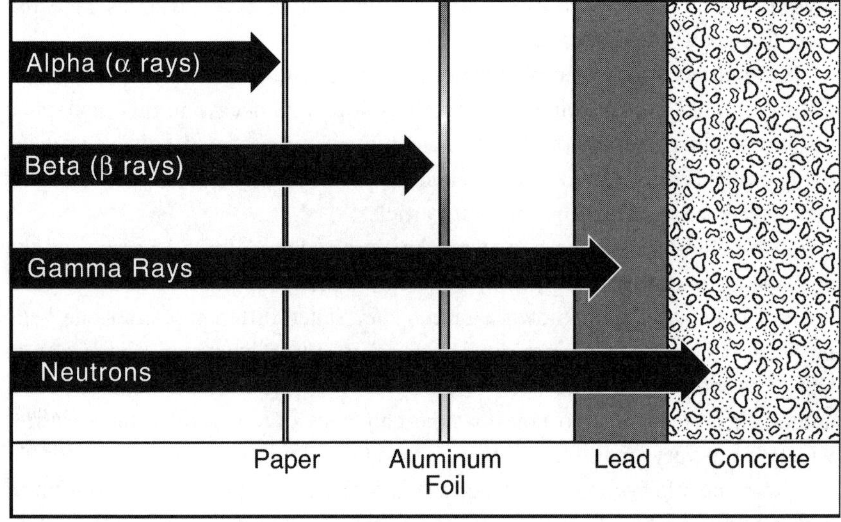

Figure 13. Shielding required to stop various forms of radiation.

Beta particles are nothing more than rapidly moving electrons. Being very light, traveling very fast, and having only half the electrical charge of alpha particles, beta particles have more penetrating power. Even so, a thin sheet of plastic or aluminum foil provides effective shielding. Positrons travel a comparable distance, soon encountering an electron in the matrix around them, at which time the two particles are annihilated (they literally disappear), being replaced by two energetic gamma rays traveling in precisely opposite directions.

Gamma rays, having no electrical charge or mass, possess a much higher penetrating power. Depending on their energy level, several inches of lead or several feet of concrete may be required to provide adequate shielding. Stopping a gamma ray in matter is somewhat complex.[1] Gamma rays lose their energy in discontinuous steps, rather than continuously as in the case of charged particles. There are three fundamentally different ways that this can be done, and all involve the gamma ray either ejecting or creating electrons within the material with which they interact. The first is the "photoelectric" effect, initially explained by Albert Einstein, where a gamma ray is absorbed by an atom in the material and an electron is ejected with the gamma ray's energy (minus a so-called binding energy). The second is "Compton scattering," where only part of the gamma energy is absorbed by the atom in the matrix—so that a lower-energy gamma ray is created (moving in a different direction), along with an ejected electron. The third is called "pair production," where the gamma ray packet of energy is converted into mass, yielding an electron and a positron. This is precisely the opposite process noted above, where a positron disappears upon colliding with an electron, yielding two gamma rays.

X-rays are very similar in nature to gamma rays, but generally have much lower energy. The crucial difference is that gamma rays emanate from the nucleus of an atom, whereas x-rays originate from outside the nucleus. X-rays are produced whenever electrons move from an outer shell to a vacancy in an inner shell of an atom or when electrons are accelerated or deaccelerated—causing a phenomenon known as bremsstrahlung.

Neutrons, having one-quarter the mass of an alpha particle but no electrical charge, also have substantial penetrating capability. Thick layers of concrete, wood, or water are normally used as shields to protect against neutrons. The reason these materials are so effective in stopping neutrons is because they contain large numbers of protons. Water is simply H_2O, and the hydrogen atoms consist of a single proton in their nuclei. A neutron can be stopped dead in its tracks if it hits a proton squarely, since they are almost identical in mass.

How Do We Measure Ionizing Radiation?

Since the human senses cannot detect ionizing radiation or discern whether a material is radioactive, how do we know something is radioactive? We now have a wide variety of special instruments that can detect and measure all varieties and energy ranges of ionizing radiation with great reliability and accuracy.

There are two fundamental classes of units that have been developed to determine radiation dose. One addresses the amount of ionizing radiation (or *dose*) a person receives. Dose is measured in terms of the energy absorbed in the body tissue. The newly adopted unit is the gray, where one gray (Gy) is one joule deposited per kilogram of mass.* However, the original unit for this measure-

* The energy in 1 Gy of radiation is actually very small. A 150-pound person subjected to 1 Gy of whole-body radiation would experience a temperature rise of only about 0.01 degree Fahrenheit.

ment is the rad (radiation absorbed dose) and has become so commonly used that I shall employ that term throughout this book. For translation purposes, 1 Gy = 100 rads.

Another set of units has been generated with the recognition that an equal exposure to different types of radiation does not necessarily produce equal biological effects. One rad of alpha radiation, for example, will have a greater biological effect than one rad of beta radiation.* Hence, when we are interested in radiation effects on the human body, we express the radiation as *effective dose*. The international unit is the sievert (Sv), whereas the more commonly used unit is the rem (radiation equivalent man). Again, for translation purposes, 1 Sv = 100 rem. Since most of our interest is focused on much lower effective doses, we will usually use the unit "millirem," or mrem, which is one-thousandth of a rem (i.e., 0.001 rem).

It is important to remember that the effect of one mrem of radiation is the same, regardless of whether the radiation is natural or created. Some people believe that a mrem of natural radiation is somehow different, or safer, than a mrem from a nuclear power plant. Not so. The human body cannot differentiate.

WHERE DOES RADIATION COME FROM?

As implied above, radiation is an innate part of nature. There is no such thing as a "radiation-free" environment. Natural radioactive material exists in the ground, in all food and water, in the air, and in every corner of the known universe—including our own bodies.

Figure 14 provides a sampling of average radiation doses of a typical American from a variety of sources. These are categorized as natural, artificial, and elective.

We note from this figure that by far the largest amount of radiation is from radon gas. Radon is a naturally radioactive gaseous material that is part of the radioactive decay chain of uranium—a very heavy element that naturally exists in trace quantities throughout the surface of Earth. This radon gas continuously seeps out of the ground, plaster, and other building materials. The amount of uranium, from which radon originates, varies considerably within the surface materials of Earth's crust. Hence, the amount of radiation exposure from radon gas

* The reason that a dose of alpha radiation is more damaging to human flesh than the same dose of beta radiation is because its energy is deposited in a very localized region, thereby causing substantial cellular disruption. On the other hand, this is only a concern if alpha radiation is inhaled or ingested, since it cannot penetrate the outer dead layer of skin if the exposure is external to the body.

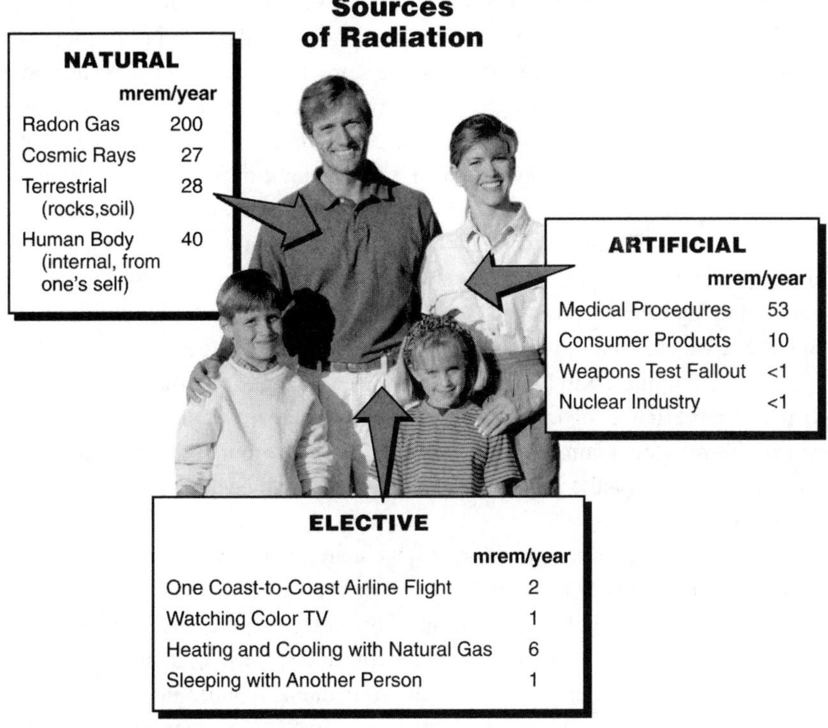

Figure 14. Average annual radiation exposure of a typical American.

varies appreciably around the country and around the globe. The 200 mrem per year attributed to radon in this figure represents a US-wide average.

Cosmic rays bathe us constantly from outer space. Cosmic rays consist of a plethora of subatomic particles that emanate from deep space—some arriving at the Earth's surface with very high energy and others with much lower energy because of secondary collisions with substances in our atmosphere. Again, the number 27 mrem simply represents a national yearly average of our dose from cosmic radiation. The principal variable is the altitude where we live, since the atmosphere is a barrier to incoming cosmic rays. The closer we are to sea level, the greater the effectiveness of the atmosphere as a shield. Annual radiation exposure from cosmic rays increases by about 1 mrem with each one-hundred-foot increase in elevation. Therefore, someone whose home is on a hill five hundred feet higher than his neighbor receives about 5 mrem more radiation every year. This means that inhabitants of the mile-high city of Denver receive about 50 mrem more radiation per year than those in Los Angeles or New York (though likely considerably less smog!). Airline crews who fly an average of sixty hours per month receive about 25 mrem of additional radiation per year because of the reduced atmospheric shielding of incoming cosmic rays at flying altitudes.

Radiation also naturally occurs from rocks, soil, and other surface materials. This terrestrial source contributes about the same yearly radiation dose as that of cosmic radiation. Radiation even comes from within our own bodies. This is mainly the result of potassium-40 in our bloodstream. The radioactive isotope potassium-40 constitutes only 1 part in 8,550 nonradioactive potassium atoms, but it is present in this small amount in all of the natural potassium element necessary for sustaining life. Essentially all of the food that we eat has at least trace amounts of radioactivity. Bananas, which contain a considerable amount of potassium (and hence postassium-40), are among the most radioactive of the foods that we eat on a routine basis, yet we suffer no ill effects.

Before moving on to the elective sources of radiation, it may be instructive to note the variation of radiation doses that occurs in regions throughout the world. Figure 15 contains a graph of annual doses from natural sources of radiation at selected global points. The variation is noted to be quite substantial—and it is all natural!

As we again refer back to figure 14, we note that the largest radiation source next to "natural background" radiation is derived from medical procedures. A chest x-ray contributes between 10 and 50 mrem, while a CAT scan may add more than 1,000 mrem. Other consumer products, such as household smoke detectors and the older Coleman lantern mantels, contribute much smaller amounts (an aggregate of about 10 mrem per year). However, even they contribute over ten times more radiation than the residual atmospheric fallout remaining from the 1960s nuclear weapons testing (now spread fairly uniformly around the globe) or from the commercial nuclear energy industry.

Finally, included is a general category of nonmedical elective sources. These include the detectable doses of radiation that we receive when we take an airline flight, watch color television, or cook with natural gas (the concern here is entrapped radon gas). Even sleeping with another person poses a radiation "issue," because every person on Earth is naturally radioactive.

HOW HARMFUL IS IONIZING RADIATION?

Given that we are all exposed to radiation every day of our lives, what do we know about its health effects on our bodies? Whereas much scientific study has been devoted to this topic, it has been difficult to get scientists to agree on the specifics of absolute health effects.

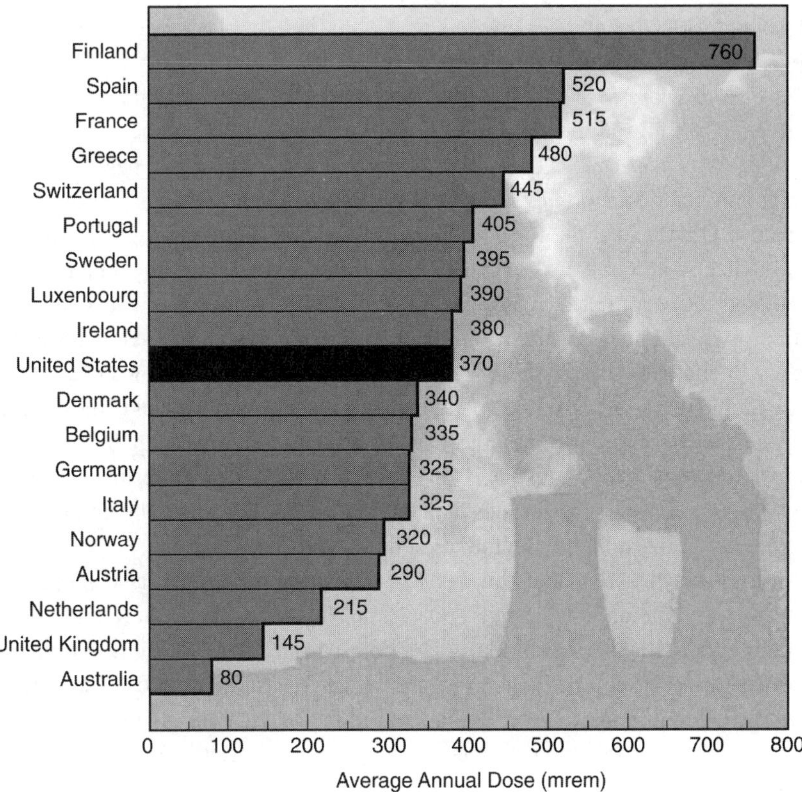

Figure 15. Average annual doses from natural radiation sources.[2]

The underlying reason for the continuing controversy surrounding this question is that scientists really don't know the precise level of radiation where there is a net harmful effect on the body. I use the word "net" here advisedly, as we shall see later.

The best data available on the incidence of leukemia (one of the most prevalent forms of radiation-induced cancer) are associated with Japanese atomic bomb survivors. We know that doses of about 500,000 mrem, delivered over a very short time span, produce leukemia in about 20 percent of people exposed to such levels. There is widespread scientific consensus that radiation levels this high and higher can be quite damaging to the human body. However, below levels of about 100,000 mrem, there is considerable controversy—especially if the dose is delivered over a relatively long time.

The lowest dose level for which reliable data has produced observable adverse health effects is around 100,000 mrem, where the leukemia rate is about 2 percent. But what are the dangers of radiation-induced cancer at the even-lower levels of say 10,000 or 1,000 mrem?

Before addressing the question, we should be reminded that a very small percentage of all cancers are actually caused by radiation. *In fact, less than 1 percent of all cancer cases are related to ionizing radiation in any way.* Food consumption patterns, smoking, and heredity traits account for about 80 percent of cancers. Leukemia is the form of cancer most likely to be induced by radiation, and radiation accounts for considerably less than 10 percent of the total leukemia incidence. A wealth of data now exists for humans exposed to low levels of radiation. It should provide some comfort to know that there are no known scientifically documented cases where low levels of radiation (i.e., below about 10,000 mrem) have caused any detrimental health effects. Nonetheless, most health physicists conservatively assume a straight line, or linear hypothesis, health effect relationship, as illustrated in figure 16. This theory presumes that any amount of radiation is hazardous, and the amount of damage to the human tissue is directly proportional to the radiation dose level.

It is worth noting that this conservative posture was initially adopted by national and international groups such as the National Council on Radiation Protection (NCRP) and the National Academy of Sciences Advisory Committee on the Biological Effects of Ionizing Radiation (BEIR) in the United States, and the International Commission on Radiation Protection (ICRP). As more and more data became available indicating no observed adverse health effects at low levels of radiation, pressure has begun to mount within the scientific community to alter this approach. Many scientists are now arguing that unnecessarily restrictive radiation standards are costing taxpayers huge sums (likely into the billions of dollars per year) for safeguards that do not have a bearing on public health or the environment.

In addition to the linear method, there are three other ways to project from the known adverse health effects of high dose levels (the Japanese atomic bomb data base) down to low levels. Curve A of figure 16 is based on a threshold effect. This approach assumes that low levels of radiation below some threshold result in very little, if any, net damage. Curve B assumes that low levels of radiation can even be beneficial (the hormesis effect). Curve C, on the other hand, assumes that low levels of radiation are relatively more harmful, on a per-unit basis, than high doses.

There is substantial scientific opinion that favors Curve A. For a detailed and easily understood treatment of the basis for this position, an excellent reference is Bernard Cohen's masterful book *Before It's Too Late.*[4] Another excellent

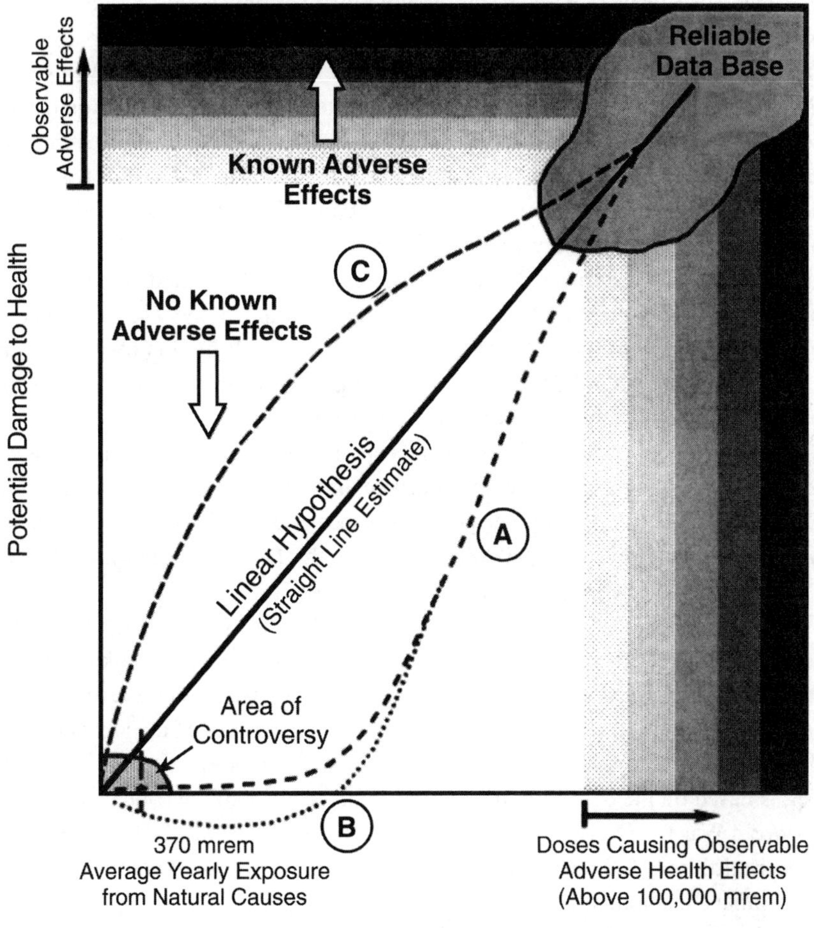

Figure 16. Models for determining the human health effects of radiation dose.[3]

source of information is Merrill Eisenbud's *Environmental Radioactivity from Natural, Industrial, and Military Sources.*[5]

Numerous studies support the scientists who subscribe to a threshold-type damage relationship as depicted by Curve A. These include widely accepted theories on how radiation induces cancer, studies of chromosome damage by radiation in human white blood cells, and experiments with mice and dogs. To be on the safe side, however, the scientists responsible for setting international radiation guidelines for industrial safety have deliberately chosen a conservative approach—that is, to overestimate the risks. They have based such guidelines on the linear hypotheses; namely, the assumption that any level of radiation, no matter how small, contributes to the risk of cancer.

It is hard to argue that zero damage is done to the human body when struck by an alpha, beta, or gamma ray. There is a finite probability that a chromosome could be damaged and some specific type of cancer could result. However, if we are concerned about the damage that may be inflicted on a vital cell or tissue, we need to recognize that the human body has about 10^{14} (100 trillion) cells.[6] Each one of these cells routinely undergoes about twenty-five thousand DNA alterations each day just from the action of free radicals* created in the normal process of combining food and oxygen to power the body. *This amounts to approximately ten million alterations* per cell *per year*. Why don't we all die an early death from such damage? It is because all living cells, including the cells of our own bodies, have exceptionally effective repair systems.

To put radiation damage into perspective, the background radiation level received by a typical human being translates into a damage level of about four DNA alterations in each cell per year.[7] This means that daily background radiation contributes only about 0.00004 percent of the DNA damage routinely suffered in our bodies owing to other natural causes.

A revelation of perhaps even more gripping dimensions is to look at comparable numbers for the health effects of high levels of radiation. We know that a radiation dose of about fifteen hundred times that of background levels (i.e., ~ 500,000 mrem) will, if delivered as a whole-body dose over a short period of time, produce a 20 percent chance of death. This observation, as pointed out earlier, is based on the data from the atomic bombs detonated over Japan at the end of World War II. Even at these exceptionally high (lethal) levels, however, radiation inflicts only 0.06 percent of the damage to our cells that is routinely caused by cellular metabolism (1500 x 0.00004). What is going on?

The only logical conclusion we can draw is that even at very high doses of radiation, it must not be the damage from the radiation that causes death. Rather, it is more likely that the repair systems become impaired at these high levels of radiation. Many medical researchers now believe that at low levels of radiation, these same repair systems are stimulated—in much the same manner as moderate exercise stimulates healthy processes.[8]

There are many substances that can be fatal when taken in large doses, yet are beneficial in small doses. Many vitamins, vaccines, and other medications fall into this category. Aspirin is a classic example. An overdose of aspirin can kill us. Yet many of us take this "modern miracle drug" routinely in low doses

---

* Free radicals are molecular fragments that have been broken off from their fully developed molecular structure. They normally carry an electrically negative charge.

for its health benefits. I do. Medical literature continues to suggest that one aspirin every day or two can have significant value in warding off heart attacks as well as providing other beneficial attributes, although it can result in stomach ulcers in some people.

Let's suppose the pharmaceutical community were to adopt the linear hypothesis approach for the health effects associated with taking aspirin. We know that taking a hundred aspirins in one sitting would likely cause us considerable harm—if not kill us. Let us assume, for the purposes of this illustration, that there is a 100 percent chance that it would be fatal. Based on the linear hypothesis theory, the practice of taking a single aspirin per day for a hundred days would prove fatal. This method doesn't recognize any benefits of time or the possibility that the body may be able to deal constructively with low doses. Yet this is the conservative theory accepted by the radiation industry for the determination of health risk assessments. Even worse, this linear hypothesis method is often applied to an entire population to determine the total "person-rems" (i.e., a dose integrated over the full population of interest) associated with a particular radiation release. This was the approach used to predict the number of potential cancer deaths resulting from the Three Mile Island and Chernobyl reactor accidents, yet the actual deaths were far less (none for Three Mile Island). Coming back to our aspirin analogy and applying it to a population of a hundred people, this method would predict one death if each person took but a single aspirin!

Until recently, medical radiation specialists in the Western world have based their studies of the effects of high radiation doses on the Hiroshima and Nagasaki atomic bomb data. However, information is beginning to surface from the nuclear weapons installations in the former Soviet Union that may offer new insights. Though preliminary in nature, the data from personnel exposed to doses in the range of 50,000 to 100,000 mrems per year (much higher than any exposure recorded in the US nuclear weapons industry) seem to indicate that there may be a significant rate effect associated with radiation damage. It appears that the human body may be able to survive very large radiation doses if the radiation is delivered over a relatively long time period—rather than suddenly, as in the case of an atomic bomb. We might parenthetically note that by far the largest numbers of deaths associated with an atomic bomb explosion are due to the blast effects, not radiation.

Radiation oncologists—cancer radiation treatment specialists—already recognize this and take advantage of this relationship. They are able to deliver more radiation to a malignant tumor with less damage to healthy cells by specifying more total treatments with a smaller dose per treatment. If careful scrutiny of this new data can confirm the rate effects, a strong basis would emerge to prove that our current practice of using the linear hypothesis is grossly conservative. The human body may be far more resilient to the effects of ionizing radiation than we previously assumed.

A recent ten-year investigation of the effects of low-level radiation on over seventy thousand shipyard workers building nuclear-powered ships is most telling.[9] It provides perhaps the best epidemiological study conducted to date for cancer and death associated with low-level radiation. This illuminating study was particularly important because it contained a large, statistically meaningful cohort of workers who all had similar working conditions, except that they were exposed to different levels of radiation at their work sites. The focus was on three specific groups who received the following work-related gamma radiation doses during the ten-year period: group 1, over 500 mrem; group 2, less than 500 mrem; and group 3, zero mrem. All workers in the study received about 3,000 to 4,000 additional mrem during this decade of observation owing to natural background radiation away from their jobs. The data conclusively showed that both groups of nuclear workers (i.e., groups 1 and 2) had a *lower* death rate from leukemia and lymphatic cancers than the nonnuclear workers (group 3). In fact, those receiving the higher radiation doses had a death rate 24 percent *lower* than those who received no radiation beyond normal background.

Other evidence is beginning to mount that, as is the case with aspirin, the human body actually benefits from a certain amount of low-level radiation. The principle for this assertion is called hormesis, shown as Curve B in figure 16. Dr. T. D. Luckey, a professor of chemistry at the University of Missouri, has presented impressive evidence for such an assertion in his landmark book *Hormesis with Ionizing Radiation,*[10] along with his updated book.[11] He bases his thesis—that low levels of radiation are actually beneficial—on over a thousand separate studies with plants and animals. James Muckerheide has compiled substantial evidence to support the hormesis basis for understanding the human health effects of ionizing radiation.[12]

This hormesis effect could help explain some of the strange phenomena that have baffled epidemiologists (medical experts who study disease control) for decades. The radioactive rock in Kerala, India, for instance, generates a radiation exposure level ten times the US average. Yet residents are reported to have the best health status in all of India, despite only average health care and the least adequate diet in the country. Numerous cases of life spans over 120 years can be cited, and the one thing these people have in common is that they live in regions with high natural background radiation (for example, the Hunzan Himalayas in Asia or the Vilcandra Andes in South America). Here in the United States, Colorado and Wyoming are both well above the national background radiation average, yet they are far below the national average in their incidence of cancer. Indeed, of the many epidemiological studies of people living in high background radiation areas, there has never been an instance of a demonstrated increase in cancer mortality.

Certainly we cannot overlook the potential hazards of radiation, particularly in large amounts. On the other hand, we cannot become paralyzed by a naturally occurring phenomenon that, according to all credible scientific data, has no harmful effect at low levels. Indeed, there is mounting evidence that we may not be able to live a healthy life without it.

Given this background, let's now see how modern science has found ways to exploit ionizing radiation and make it work for us.

3. Harnessing Radiation

Now that we know more about radiation, let us explore how we can safely harness it to serve humanity.

Our first step is to understand and evaluate its unique characteristics. This requires defining the "hooks" upon which to fasten the harness. Our forefathers have known for centuries that oxen and horses possess strength far beyond that of humans. Consequently, they built harnesses to translate their energy into pulling plows and to perform other useful work. Chemical energy, originally exploited in campfires for heating caves and cooking meals, has been recognized to have explosive characteristics when certain fuels are mixed. This "hook" gave rise to the internal combustion engine—allowing tractors of enormous "horsepower" to considerably increase efficiency on the farm. This same internal combustion engine continues to be the primary source of power widely used today for cars, trucks, ships, and locomotives. Early observations that electrons could be directed to move in a preferred direction by applying a voltage to one end of a conductor has given rise to vacuum tubes and then computer chips capable of transmitting almost unfathomable amounts of data in modern computer systems.

As we now look to radiation, we find that there are over a dozen unique aspects of the radioactive decay process that can be and have been tapped and harnessed for beneficial human use.

1. Material penetration—As noted earlier, different types of particles emitted during radioactive decay have unique material penetration powers. For instance, if a beam of electrons (beta particles) of a known energy is focused upon a thin sheet of aluminum, the precise thickness of that sheet can be determined by measuring the reduction in the beam current on the other side of the metal. This property (also referred to as attenuation) is used in countless commercial operations such as making rolled metals or paper, filling containers, and so on.

The most commonly recognized application of radiation penetration, and associated attenuation, is for imaging. A perfect example is an ordinary dental x-ray. Here the dentist focuses an x-ray beam on one side of your mouth and places a film on the other side of the tooth. That's the little "wad" that he or she asks you to bite down on and then cautions you to hold still. As the x-ray traverses the area of the tooth, it "zings" right through the gum and exposes the film. But the tooth is more massive, and very little of the x-ray can penetrate. Hence, that area of the film is not exposed. Should there be a cavity in the vicinity, a bit more of the x-ray can get through to the film. Hence, a "gray" spot appears at that point—and you are then invited to fork over a tidy sum to finance a filling.

X-rays are also commonly used to check the health of your lungs (chest x-rays), as well as to assist the orthopedist should you have the misfortune of breaking a bone somewhere in your body. The amount of radiation that you normally receive from such procedures is now quite small (a few mrem) and certainly worth it. Most people would not want the dentist to pull the wrong tooth or to have a broken bone set improperly.

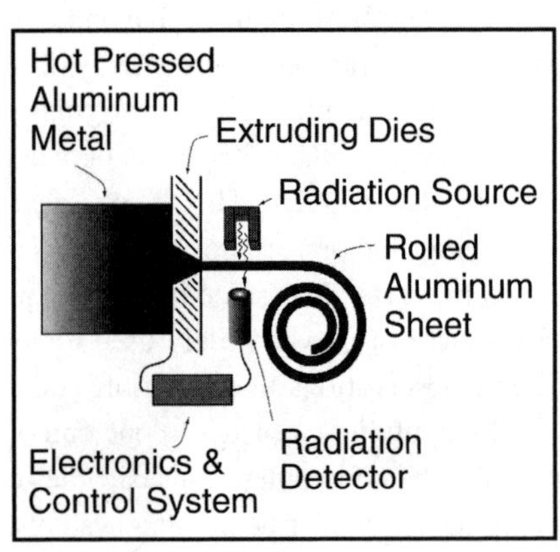

Figure 17. Material penetration.

2. *Heat source*—Alpha and beta particles are stopped within short distances of their origin as they collide with atoms in their encasement material. Substantial heat is generated in this process, much like the heat generated in a skidding tire on dry pavement. This heat can be used to generate electricity, and it is precisely this technique that is used to power deep space probes. The vivid pictures returned to Earth from Mars during the *Viking* mission were made possible by radioactive heat sources. This is the technique utilized to power the *Voyager* "twins" that we were introduced to in the opening chapter. Properly designed, such power sources can last for decades—without any moving parts.

3. *Emission*—Many isotopes emit gamma rays during the radioactive decay process, often in conjunction with other particle emissions. Since gamma rays can penetrate through a relatively large mass, it is possible to attach specialty gamma-

emitters to materials via chemical means that can easily be transported by fluids. Such a linkage allows the gamma-emitting radioisotopes to flow through a set of pipelines or blood vessels, thus allowing the location of this radioactive material to be measured as a function of time. This *tracer* or *labeling* technique is routinely used to map groundwater movement, detect pipe leaks in the petroleum industry, and diagnose ailments in the human body. As might be expected, this technique has made a powerful impact on the world of medicine and, in fact, is precisely what was used confidently to convey that my mother did not have bone cancer.

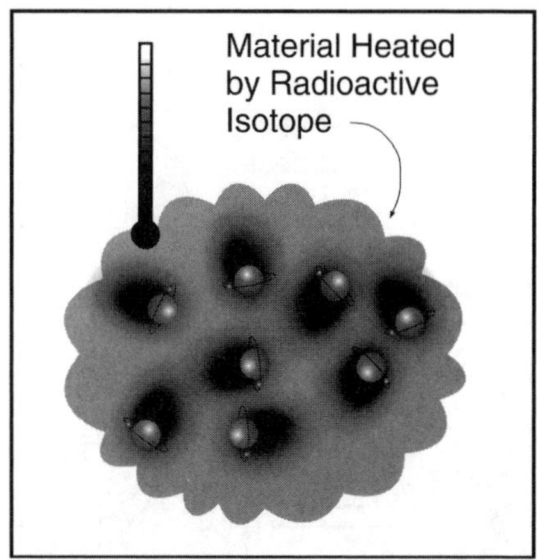

Figure 18. Heat source.

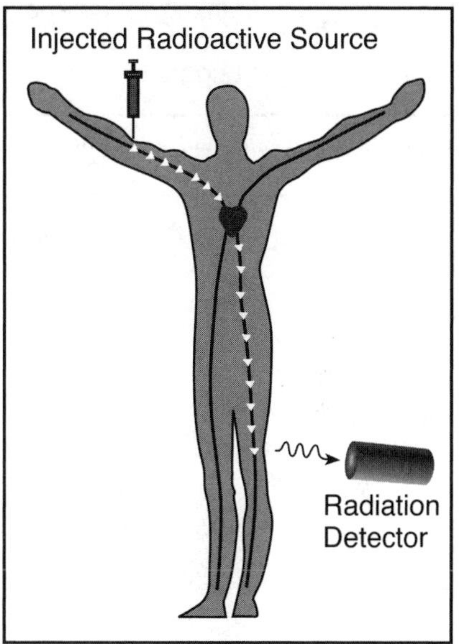

Figure 19. Emission.

4. Transmutation/activation—By focusing a beam of neutrons on unknown materials, it is possible to transmute (fundamentally change) the target material into another isotope (or possibly an isotope of a different element). If this new isotope is radioactive, it will emit a radioactive signature that uniquely determines its identity. It is then a very simple process to identify the target material (the original unknown material) that was converted to the measured resulting product created by absorption of the neutron beam. This property is used routinely for criminology investigations and for revealing explosives in airport luggage.

Living after the 9/11 terrorist attacks, I find it comforting that essentially any type of explosive material can now be detected by such techniques in scrutinizing baggage. It is possible that this activation technique was used when that friendly luggage inspector at the airport took a small piece of cloth to "swipe" around your briefcase or tote bag and then placed the cloth into a special box for analysis.

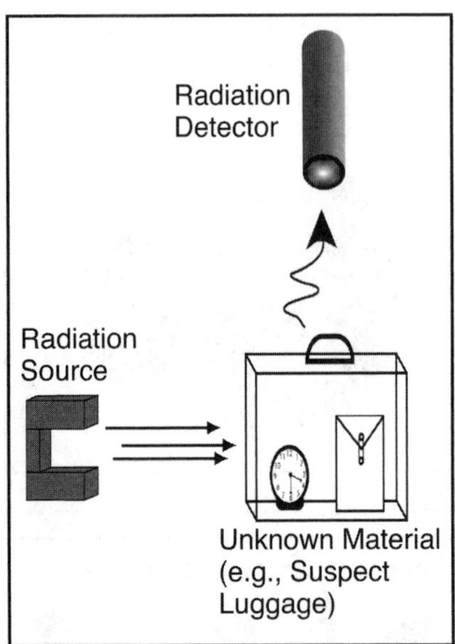

Figure 20. Transmutation/activation.

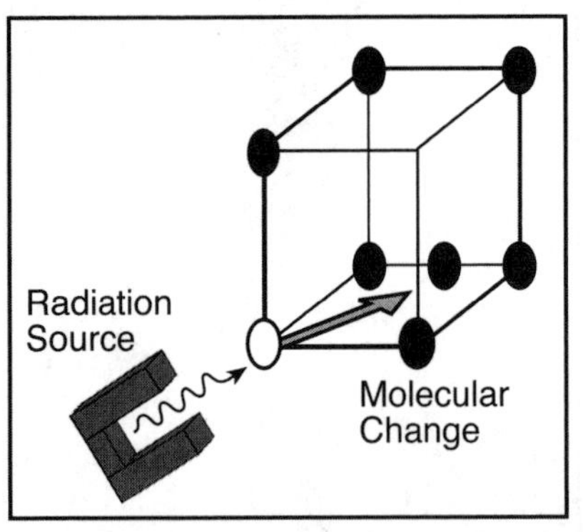

Figure 21. Change chemical structure.

5. Change chemical structure—Bombarding some materials with intense beams of gamma rays, electrons, or neutrons can break molecular bonds and evolve new types of materials (i.e., change the chemical or physical structure). Such treatment is used to produce stronger plastics or increase the rate of a chemical reaction. This approach is even used to enrich the color and brilliance of some types of gemstones by slightly modifying their structure.

6. Cell destruction—In a manner similar to the above, controlled amounts of beta particles, x-rays, or gamma rays can be used to kill unwanted insects or microorganisms. This property is used in millions of operations per year for sterilizing hospital equipment, killing food pathogens (i.e., any microorganisms that can cause disease), suppressing sprouting in vegetables, and eradicating cancer cells. As hinted at in our opening story, it is this property that allows for a potential transformation in our massive food industry—to provide safer produce with a longer shelf life.

7. Decay time—Knowing the "half-lives" of certain radioactive species (i.e., the time it takes for one-half of the atoms of a certain species to decay into another species), it is possible to perform aging analyses on a variety of archeological artifacts—including the age of Earth. This is the technique used to determine prehistorical climate changes, the role of the industrial age in contributing to greenhouse gases, and so on. The list is endless.

8. Luminescence—Particles emitted from radioactive decay can cause the emission of visible light when interacting with certain target materials. Tritium is an example of such a radioactive material that can be incorporated into materials that luminesce. It is routinely used in the form of a "paint" for lighting airport runways, exit signs, or other safety equipment in remote areas where it is not economical to deliver electricity or where 100 percent reliability must be guaranteed. Recall our airport story in Quito, Ecuador.

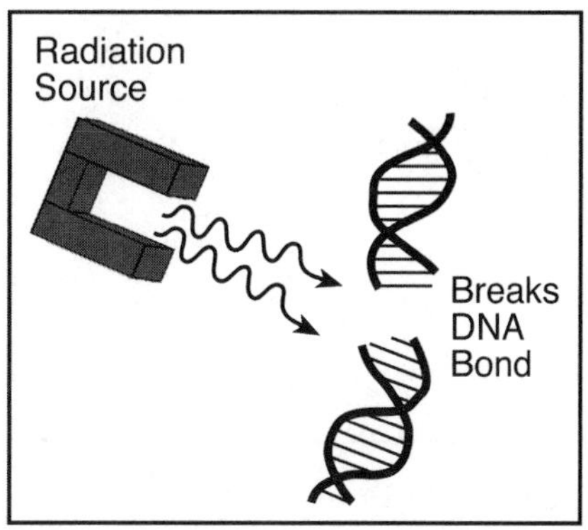

Figure 22. Cell destruction.

9. Ionization—Radiation striking outer-shell electrons can, if sufficiently energetic, knock such electrons out of orbit and cause the targeted material to become ionized. As we discussed earlier, this is called ionizing radiation. The targeted material then contains free radicals. When the striking radiation is very intense, it can induce polymer chains to form useful products such as polyethylene and heat-shrink plastics.

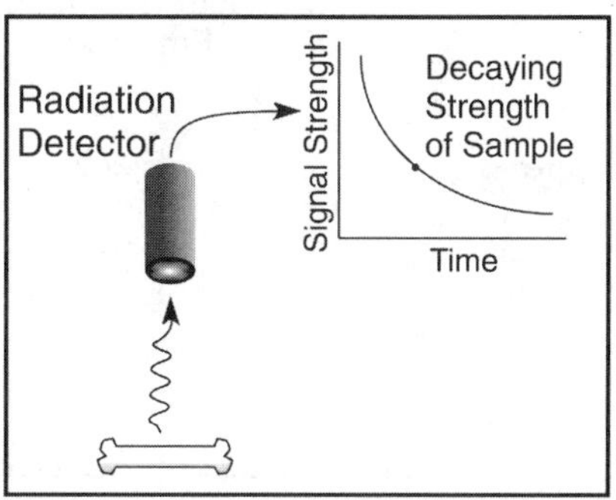

Figure 23. Decay time.

10. X-Ray fluorescence—X-rays focused on a material of unknown composition can knock out inner electrons from the atomic structure of the unknown material. When this happens, electrons from the outer shells will refill the vacated site. The movement of these electrons causes the emission of x-rays with energies and relative intensities characteristic of each element comprising the unknown material—thus allowing precise identification of the unknown composition. This is a powerful method for measuring concentrations of trace elements and is used in any fully equipped analytical laboratory. It is sometimes used for criminology work.

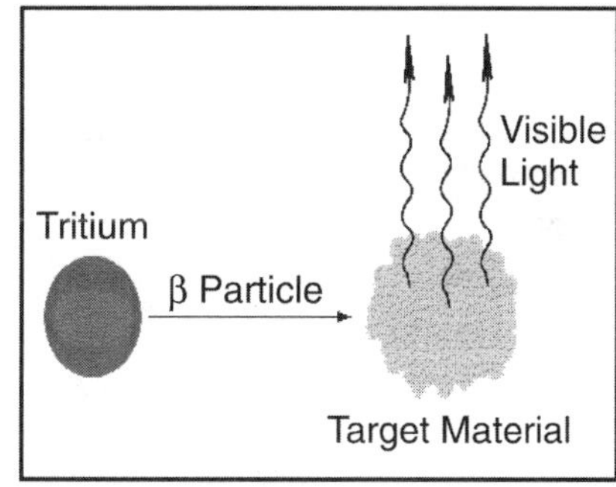

Figure 24. Luminescence.

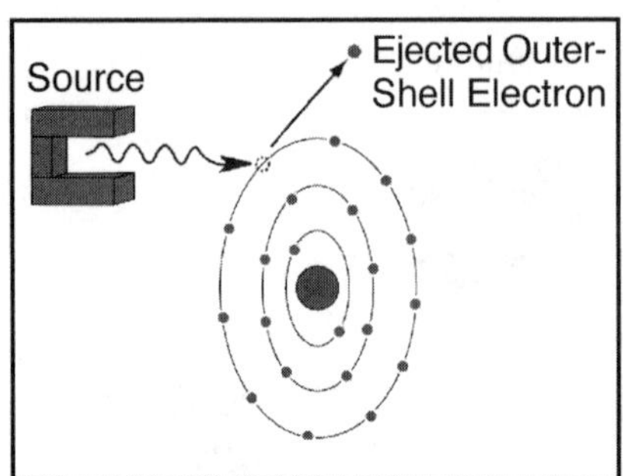

Figure 25. Ionization.

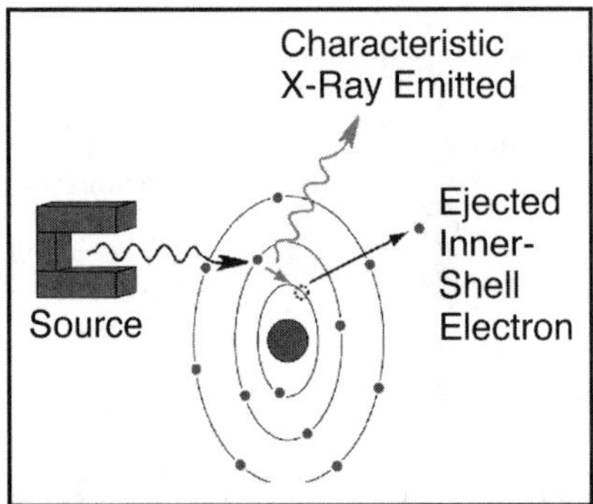

Figure 26. X-ray fluorescence.

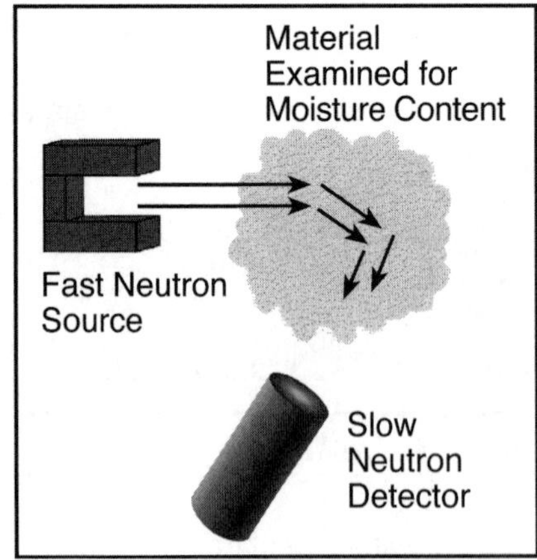

Figure 27. Neutron slowing down.

11. Neutron slowing down—If a burst of energetic (fast) neutrons is introduced into a media, these neutrons will collide with the surrounding atoms—thus losing energy and slowing down to velocities typical of ordinary materials (i.e., thermal speeds). This slowing-down process can take place very rapidly if there is substantial water in the surrounding media, since water contains hydrogen (an element consisting of a single proton). A fast neutron colliding with a proton can lose all of its energy in a single collision—in the same manner as a cue ball that is stopped in its tracks upon a "bull's-eye" collision with the eight-ball. However, when a fast neutron collides with a heavier element, such as iron or lead, it loses its energy gradually and many collisions are required for the neutron to reach thermal speeds. This property can be used to obtain an accurate measurement of how much moisture is contained in a particular sample (such as soil), since the number of slow (thermal) neutrons detected is directly proportional to the number of water molecules in the immediate vicinity.

12. Neutron or photon scattering—As a variation to the slowing-down property noted above, if a neutron colliding with a nucleus of unknown identity does not make a "bull's-eye" hit, it will scatter forward or backward at some angle relative to its original direction. This information can be used to determine the identity of the unknown material. Gamma or x-ray backscattering is a variation of the same theme. The difference between gamma or x-ray scattering and neutron scattering is that gamma and x-rays interact with the atomic electrons whereas the neutrons interact with the nuclei of the atoms. These techniques can be used to analyze materials on the surface of bulk matter, such as the thickness of plastic coatings on metals.

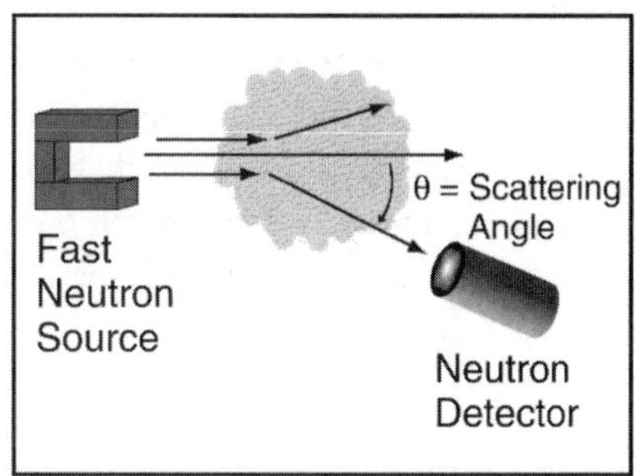

Figure 28. Neutron or photon scattering.

13. Fission—For certain heavy isotopes such as uranium-235 or plutonium-239, the absorption of a neutron can cause a transformation to take place that produces a nucleus sufficiently unstable that it will literally break apart, generally resulting in the release of two fission fragments, several new neutrons, and an enormous amount of energy. This is the basic process that enables nuclear fission reactors to operate and generate large amounts of electricity.

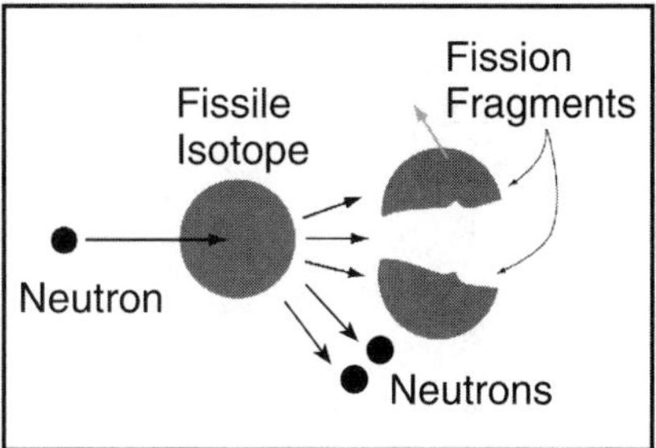

Figure 29. Fission.

14. Fusion—For very light isotopes, such as deuterium (a hydrogen atom containing a neutron in addition to a proton) or tritium (a hydrogen atom containing two neutrons in addition to a proton), it is possible that a collision of deuterium or tritium at high enough speeds (i.e., exceedingly high temperatures!) will cause the two nuclei to fuse and form a new isotope—yielding an energy release that is even greater than in the fission process outlined above. This process is called nuclear fusion. It is the energy source for our sun and all the other stars. Fusion may eventually be harnessed to produce electricity, but many difficult scientific and engineering challenges remain to be resolved before this becomes a reality.

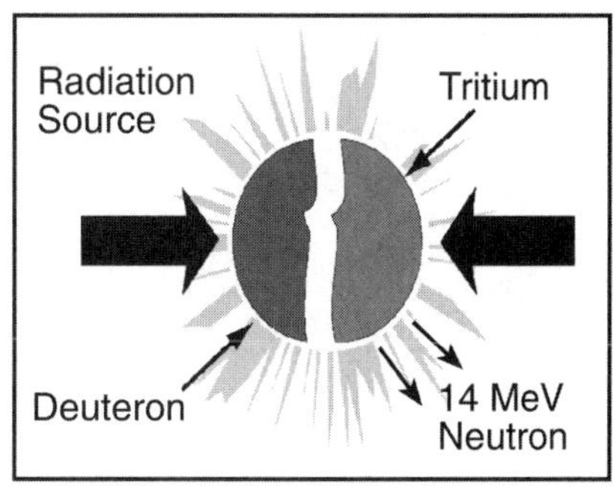

Figure 30. Fusion.

It is my hope that this brief overview has helped take some of the mystery out of how radiation can be productively employed. Its many unique characteristics afford a plethora of opportunities for scientists and engineers to tap its power for beneficial use. Now that we have a better understanding of where to "strap the harness," we can begin our investigation of the numerous ways that radiation has so improved our lives.

4. AGRICULTURE

There is perhaps nothing more fundamental to human survival than having access to adequate and safe food supplies. As far back as the beginning of recorded history, and likely eons earlier, humankind has struggled to find sufficient food to stay alive. We might be tempted to believe that there were periods in our early history when people found food foraging to be easier than in modern times, since the population density was low enough to "live off the fat of the land." In actuality, any factual reflections clearly reveal that a much larger fraction of the workday was spent on acquiring basic food supplies than is the case in a modern, robust economy.

Our struggle today and in the future is that population numbers have exploded within the past century, with billions of more mouths to feed than ever before. Sad to say, around 850 million of our fellow citizens on planet Earth (approximately one out of every seven) go to bed hungry every night. Tens of thousands die every day from hunger and hunger-related diseases. Hence, there is an enormous need to find new ways to increase food production and deliver it with only minimum spoilage to an increasing population of hungry mouths.

We recognize that political and cultural environments also factor heavily into the availability of food supplies to world citizens. However, it is comforting to know that radiation processes are now available and widely used to improve the productivity of the entire food chain in a substantial manner.

Before delving into the specifics, however, I'd like to share an amusing story about how a rather ingenious young lad found a way to use radiation over ninety years ago to do a bit of food-quality sleuthing on his landlady.[1] The setting was

1911 in Manchester, England, where George de Hevesy was a struggling Hungarian student forced to live on a very low budget. As such, he had all of his meals prepared by the landlady who provided his modest accommodations. Though not one to be ungrateful, he began to suspect that some of his meals might have been made from leftovers from the preceding days—or even weeks. To confirm his suspicions, he put a small amount of radioactive material into the leftovers of one of his meals. Several days later, when the same dish was served, he used a simple radiation detection instrument (a gold leaf electroscope) to check the food. With mixed emotions, he observed a slight deflection in the gold leaf and confirmed the existence of leftovers. (Don't try this at home, kids!).

George's landlady was soon forgotten, but our budding young "investigator" went on to pursue a brilliant career in science. This culminated in Dr. de Hevesy's winning a Nobel Prize in Chemistry in 1943 and the Atoms for Peace Award in 1959. Who would have thought that a clever manipulation of leftovers might have inspired such an accomplished future?

But now back to the present. We certainly don't want to put radioactive material into our food like the young inquisitor above. In fact, that is precisely what we do *not* want to do! Rather, modern science has found ways to employ radioactivity to provide higher crop production, improve animal productivity, eradicate pests, and process the final food product to enhance its shelf life and make it safer for human consumption. The results are quite impressive.

HIGHER CROP PRODUCTION

It is well known that the yield from any crop is directly dependent upon proper amounts of nutrients and water. The demand for fertilizer, a necessity in modern agricultural practices to maximize crop yields, will continue to mount in order to provide food for the rapidly increasing world population. Radioisotopes are being used to "label" different fertilizers by employing the particle-emission (or tracer) property. Radioactive tracers are attached to different kinds of fertilizers, allowing scientists to determine associated nutritional deficiencies as the labeled products are absorbed at critical locations in the plant. This technique is used to substantially reduce the amount of fertilizer required to produce robust yields, thereby lowering costs to the farmer and minimizing environmental damage. This also avoids the need for substantial amounts of energy consumption, since the fertilizer industry is highly energy intensive.

Water is likewise an essential factor for crop production, and it is becoming critically scarce in many areas of the world. Neutron moisture gauges, operating on the principle of the slowing down of neutrons, are used to measure the hydrogen component of water in both the plant and the surrounding soil—thereby determining whether there is sufficient moisture for optimal plant growth. They are ideal instruments to help farmers make the best use of limited water supplies, preserving water for other uses. Energy costs required for pumping are also reduced. By using only the minimum amount of water that is actually needed, this practice also minimizes the runoff of potential contaminants into nearby lakes, streams, and rivers.

Another effective way to improve crop production is the development of new varieties: varieties that can better withstand heat or storm damage; exhibit enhanced maturing times (to escape frost damage and allow crop rotation); attain increased disease and drought resistance; provide better growth and yield patterns; deliver improved nutritional value; allow improved processing quality; and provide color or other attributes to enhance consumer appeal. For centuries, selective breeding and natural evolution resulting from spontaneous mutation (i.e., randomly occurring changes in the plant's genetic makeup) have winnowed out weak plant types, allowing only the hardiest to survive. But specialized radiation techniques (either directly irradiating seeds to alter DNA structures or irradiating crops to induce variations in the resulting seeds) can greatly accelerate this process to produce superior varieties.[2] Subjecting plants and seeds to carefully tailored ionizing radiation in order to create new combinations in their genetic makeup has resulted in improved strains of such staples as barley, millet, sorghum, wheat, rice, maize, garlic, bananas, beans, avocados, and peppers. Other crops include new varieties of cotton, soybeans, sunflowers, and peppermint. New fruit varieties have been developed in apples, cherries, oranges, peaches, bananas, apricots, pomegranates, pears, and grapefruit. There is even one cultivar (mutant variety) for raspberries and grapes. Of the 2,250 new crop varieties, 75 percent are crops of the type just mentioned. The other 25 percent are ornamental flowers, such as chrysanthemums, roses, dahlias, bougainvillea, begonias, carnations, alstroemerias, achimenes, streptocarpus, and azaleas. There may be many more varieties of flowers, but they have not been openly recorded, since commercial companies keep special varieties a trade secret. Such specialty flowers typically start with a pink color but then mutate to red, white, or yellow upon being irradiated.

Many of these superior varieties now constitute a key part of modern agricultural commerce around the world. In fact, over thirty nations have taken part in the development of the more than 2,250 new crop varieties in the past seventy years—with radiation being the key element in the development of 89 percent of this enormous new stock.[3]

Indeed, the application of radiation techniques for developing new crop varieties has likely provided the highest global economic value of any attribute of radiation. Mutant varieties (called cultivars) of numerous crops have been created via radiation, where basic alterations in plant DNA structure have been accomplished by gamma ray, x-ray, and neutron bombardment.

All of this technology started as far back as 1928 with the discovery of mutagenesis—an important tool for locating genes on chromosomes. Plant breeders and geneticists soon became interested in radiation as a fast and effective way to alter plant traits. Gamma rays are used in the majority of cases to change plant characteristics (64 percent usage), and x-rays are employed in another 22 percent of the cases. The bulk of the remaining 14 percent is done via fast and thermal neutrons. To date, China has benefited the most from the utilization of radiation to improve crop species. As of 2002, nearly 27 percent of the crops grown in China were developed using radiation techniques. China is followed by India (11.5 percent), the USSR/Russia (9.3 percent), the Netherlands (7.8 percent), the United States (5.7 percent), and Japan (5.3 percent).

Rice is the major source of food for over 50 percent of the global population and is especially important in the Asian diet. Some 434 mutant varieties of rice have been developed, of which half were developed from gamma radiation. In Egypt, semidwarf varieties developed by radiation have increased the yield to nearly nine tons per hectacre (over twice the world average). Japan has grown eighteen mutant varieties of rice, which yielded $927 million for the Japanese economy in 1997. Because of radiation, Thailand has become the largest exporter of aromatic rice in the world. During the decade from 1989 to 1998, Thailand produced $19.9 billion worth of milled rice.

Barley is a prime ingredient in making malt. Mutant varieties such as 'Diamant' and 'Golden Promise' are two radiation products that have made a major impact on the European brewing and malting industry. One of my major sources of information came from Dr. James Dargie, director of the Food and Agriculture Organization within the International Atomic Energy Agency in Vienna, Austria.[4] Being a native of Scotland, Dr. Dargie is very proud of the fact that the premium quality ales and Scotch whiskey are made from 'Golden Promise'. This industry has provided Scotland with a revenue of approximately $417 million over the last quarter century. Both the United Kingdom and Ireland likewise make wide use of 'Golden Promise' for their beers and whiskey. How many of you readers of drinking age, who enjoy a bit of alcoholic refreshment now and then, are aware of the debt of gratitude owed the radioactive atom?

Wheat is the staple grain for many countries, including the United States. Pakistan has developed three mutant wheat varieties that have provided $47 million to their farmers over a five-year period. In Italy, the Durum wheat 'Creso' mutant was developed via radiation. By 1984 this variety reached 53.3 percent of the Italian market, yielding $180 million per year. I often asked my students which country came immediately to mind when I mentioned the word *pizza*. And, of course, the resounding unanimous reply was *Italy*. Given the above statistics, it is fair to say that over 50 percent of the pizza consumed today in Italy is the direct result of harnessed radiation. Mamma mia!

Cotton has been the staple of the clothing industry in many nations for decades. A special high-yielding cotton mutant NIAB-78 was produced by gamma rays and released for commercial production in 1983. This variety has shorter stature (so that the nutrients go into the product rather than the stock), better growth, enhanced heat tolerance, and is resistant to bollworm attack because of its early maturity. It likewise provides an ideal cotton-wheat rotation planting pattern. This variety has had a pronounced impact in Pakistan, where the country's entire clothing industry was threatened by an insect infestation. Within five years of the release of this new, radiation-induced variety, the cotton production in Pakistan doubled. Within ten years of release, this variety yielded over $3 billion in cotton production.

One of the earliest applications of this radiation technique was to save the US peppermint plant.[5] Having succumbed to a fungal disease, this species was threatened with extinction. All crossbreeding efforts failed. Radiation bombardment led to a new variety able to ward off the fatal disease, thereby saving the original peppermint taste enjoyed by millions throughout the world. You might remember and appreciate this fact the next time you savor a stick of peppermint gum.

An application of radiation to crop production that particularly caught my attention was the improvement of the Texas grapefruit. For years our family has enjoyed a box of grapefruit from southern Texas. We could get ahold of this special treat while living in the state of Washington, but during our recent four years in Texas, we found it more difficult to come by. Why? Like many commodities, the best of the best is often for export only. It was more profitable to send this premium product out of state. Being the inquisitive type, I wanted to find out what made this product so special. I learned to my delight that 'Ruby Red' started the commercial grapefruit market in Texas from a mutant that originated from spontaneous mutation in 1929. It's nice to know that something good happened at the start of the Great Depression.

Well, the flesh color of 'Ruby Red' faded as the harvest season progressed, and the juice's color did not appeal to customers. A redder grapefruit was deemed more desirable. Hence, seeds and bud wood from 'Ruby Red' were irradiated with thermal neutrons, which brought two new cultivars into being. The most successful was called 'Star Ruby', released in 1970. It was a seedless variety with red flesh, and it quickly gained wide acceptance. Still, the yield of this variety was quite variable. Consequently, the bud wood was again irradiated with thermal neutrons, and the further-improved product, 'Rio Red', was released in 1984. It retained the red flesh of 'Star Ruby' and resulted in a good, reliable yield. Both varieties now provide the mainstay of the premium Texas grapefruit market—both sold under the trademark 'Rio Star'. They are grown on 75 percent of the land devoted to grapefruit production in Texas.

Having had the opportunity to travel to Japan on numerous occasions, I am also struck by the way radiation has made an important contribution to the famous Japanese pear. Long a trademark of Japanese cuisine, the original species became susceptible to blackspot disease and was threatened with extinction. Fortunately, 'Gold Nijiseeiki' was developed by using gamma rays, and it reestablished this famous pear—yielding an additional income to Japanese farmers of about $30 million per year.

The worldwide floriculture (flower) industry is estimated to be worth $70 billion a year, and it is growing rapidly, especially in the developing nations. Countries like Indonesia, Malaysia, and Thailand have targeted increases in land area for floriculture products to enhance national exports. Malaysia, for example, has released a radiation-induced mutant orchid variety, Dendrobium 'Sonia KeenaAhmadSobri', which has diamond-shaped petals and a shelf life of fifteen days. Another variety, called *Tradescantia spathcea* var. Sobril, has creamy stripes and is best used in home flower pots. Similarly, six new mutant varieties of chrysanthemum have been released in Thailand, with flowers ranging in color from yellow to white, purple, yellow orange, and orange.

The above examples only begin to describe the impact that radiation has had in the development of crops and flowers. Its results are nothing short of amazing. This is worth more than passing reflection the next time you wander down the produce aisles of your favorite supermarket. The array of produce at our fingertips, which we all too often take for granted, didn't "just happen."

IMPROVING ANIMAL PRODUCTION

Farm animals, essential for providing commodities such as milk, meat, wool, and leather, have likewise benefited from the application of radiation techniques. One key venue concerns the best way of feeding animals: whether to graze the animals on only natural pastures or to provide supplemental commercially prepared feeds. In order to answer this question, researchers labeled the feed with specialty radioisotopes, such as carbon-14, and then traced the paths of the food within the animals' digestive systems. This determines where and how quickly the food is broken down into body tissues or milk. These tracing techniques have been used to determine the nutritional value of the feed to the animals and also the nutritional value of the products, such as milk or meat.

In many parts of Asia, the primary feedstock for buffaloes and cattle consists of rice straw and native grass. This combination often lacks sufficient protein, energy, and minerals that are needed for a balanced diet. By employing tracer radioisotopes, scientists can determine key nutritional deficiencies. India and Indonesia have effectively used this approach. In one example, scientists in Indonesia were able to develop a multinutrient block for buffaloes to lick. This increased buffalo weight gain at the rate of 3 kg per week and at the same time reduced their grass consumption by 80 percent.[6] In another instance, Indian scientists used tracer techniques in developing feed supplements for cows to enhance the production of milk. This allowed the largest milk co-op in India to increase milk production by 30 percent from the same-size herd and accomplish this at a production cost 25 percent below that achieved by feed supplements that did not use radioisotopes in the development process.[7]

Controlling Insect Pests

Unwanted insects can present an annoyance under a multitude of conditions. Whereas I try hard to appreciate the supposition that all insects have a purpose in our overall ecosystem, I must admit that I am still searching for a reason for mosquitoes! I am still awaiting a meaningful explanation of the value to humanity of those bloodsucking pests.

But for farmers, unwanted insects can become far more than a mere annoyance. They can lead all the way from substantial livestock disease and loss to complete devastation of crops. Estimates of harvest losses resulting from unwanted insects range from about 10 percent annually to as high as 30 percent in some developing countries.[8] Synthetic insecticide treatments used to control such insects often destroy pollinators and other beneficial insects, while also creating environmental pollution and even toxic residues in the food chain. Furthermore, many insects have developed enough resistance to insecticides to force the use of ever-higher quantities and combinations of insecticide for them to be effective.

One proven way of suppressing or even eradicating unwanted insects is to use the nuclear technology–based sterile insect technique (SIT). This approach involves rearing a large population of the unwanted insects, subjecting the maturing insects to sufficient levels of gamma irradiation to achieve sexual sterilization (but leaving other capabilities unchanged), and then releasing the sterilized males into their native environment. When the sterile insects subsequently mate with the wild insects, guess what? No babies! In addition to being environmentally sound, this technique is often the only practical means to ensure pest eradication. And there are no secondary health hazards to humans as is the case when heavy doses of insecticides are used.

Perhaps the largest successes to date in utilizing this technique are the ones achieved in Mexico against the Medfly (Mediterranean fruit fly) and throughout North and Central America against the screwworm.[9] By 1982 essentially complete success was declared against Medfly, a pest that results in wormy fruit and vegetables. Through this continuing SIT effort along the border with Guatemala, Mexico has protected its horticultural export industry, allowing it to expand (particularly since the North American Free Trade Agreement) to over $3.5 billion per year. By 1991 the screwworm was eradicated, representing major benefits to the livestock industries, yielding some $1 billion in annual benefits to the Mexican economy. Given this success, sterile insects from the Mexican production plants were then applied to infestations of such insects in Libya and Central America.

FOOD PROCESSING

Once food has been grown, it is crucial that this precious commodity be preserved and protected against contamination until consumed by an increasingly hungry world. Tragically, infestation and spoilage prevents at least one-fourth of the world's annual food production from reaching the mouths of its citizens. The percentage of harvested seafood that never reaches a human mouth is even higher—sometimes well over 50 percent. This is particularly the case in countries with warm and humid climates, characteristic of many developing nations.

In addition to the spoilage of massive quantities of needed nourishment, food can become unsafe for consumption through contamination by bacteria, parasites, viruses, and insects. The US Centers for Disease Control and Prevention estimated in 1999 that approximately five thousand Americans die each year from food-borne diseases, beginning with symptoms including nausea, cramps, and diarrhea.[10] Furthermore, some seventy-six million US citizens become sick from food-related illnesses each year, and approximately 325,000 of these persons are hospitalized.

For the most part, these deaths occur one by one—with little public attention. But there have been several instances in the past decade or so that have led to public outcries. One such event occurred in 1993 when several children in Seattle were stricken by *E. coli* (0157:H7) traced back to undercooked hamburgers at a fast-food restaurant. Subsequent outbreaks have led to huge recalls of meat and poultry products. A large recall in the United States occurred in 1997 when Hudson Foods ordered twenty-five million pounds of hamburger patties to be destroyed.[11] In October of 2002, Pilgrim's Pride Corporation recalled twenty-seven million pounds of turkey and chicken products after forty people became very ill and eight died from *Listeria* poisoning in the Northeast.[12]

Food preservation methods have evolved from the earliest days of sun drying to salting, smoking, canning, heating, freezing, and the addition of chemicals such as methyl bromide. Fortunately, food irradiation is now positioned to provide a substantially superior method. It is becoming an increasingly popular way for fast-food restaurants to protect themselves from litigation as a result of food poisoning. Furthermore, fast-food corporations are beginning to irradiate the shredded lettuce, sliced tomatoes, and other condiments to extend their shelf lives.

What Is Food Irradiation and How Does It Work?

Food irradiation involves subjecting the food to carefully controlled amounts of ionizing radiation, such as gamma rays or highly energetic electrons, to break the DNA bonds of targeted pathogens. This approach is especially effective in destroying the reproductive cycles of bacteria and other pathogens. Such radiation can reduce or eliminate unwanted organisms and specific, non-spore-forming pathogenic microorganisms such as *Salmonella*. It can also interfere with a physiological process such as sprouting in potatoes or onions, or the ripening of some fruits. Thus, the shelf lives of many foods can be extended appreciably, and food-borne disease organisms such as *E. coli* can be dramatically reduced or eliminated.

The electrons are produced by machines (accelerators), whereas the gamma rays are normally produced by the radioactive decay of cobalt-60 (although cesium-137 is also being used on a limited scale). X-rays can also be effectively used. They are normally produced by accelerators in which the highly energetic electrons are directed onto a target material such as tungsten, which then converts the energy into x-rays via a process called bremstraalung.

During the irradiation process, prepackaged or bulk food is transported by a conveyer system into a thick-walled room that houses the irradiator. The food is exposed to carefully controlled amounts of ionizing radiation for a specific time to achieve certain desirable objectives. It is then removed from the irradiation beam and placed onto a truck for delivery to the consumer.

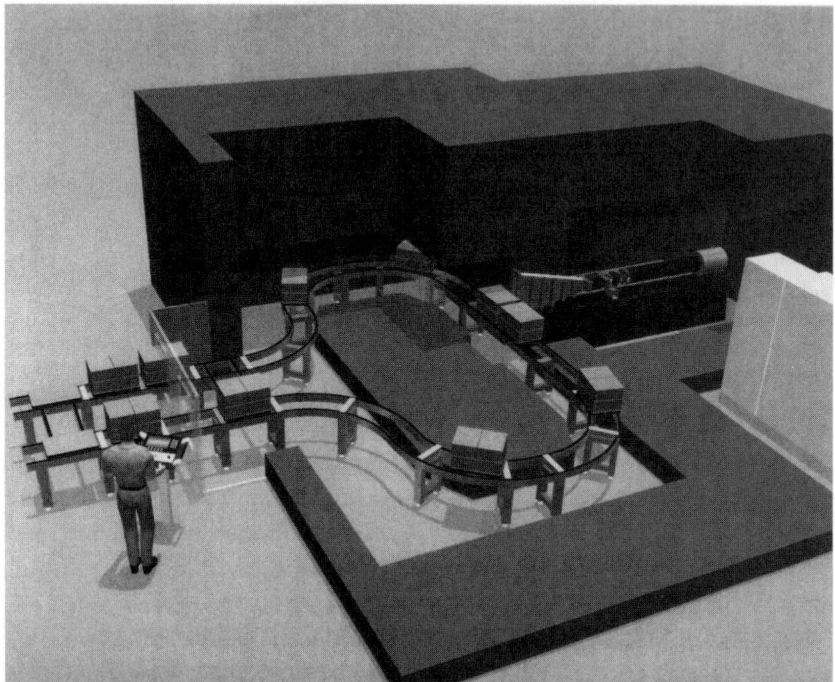

Figure 31. Schematic of a food irradiation processing facility. (Courtesy of the Institute of Food Science & Engineering, Texas A&M University.)

It is important to note that the processed food does *not* become radioactive. At the doses prescribed, it is *impossible* for the electrons, gamma rays, or x-rays to transform (transmute) the food into becoming radioactive. By law, the energy of the radiation must be less than 5 MeV, an energy insufficient to cause nuclear reactions in the irradiated food.

In the story that opened this chapter, the young George de Hevesy intentionally put some radioactive material in his leftovers, and that would, indeed, result in some radioactive material being left in the food. But in the food-irradiation process, no radioactive material is either put into the food nor does the process cause any of the food to become radioactive. In fact, it is very difficult to determine that the food has been irradiated without the use of very sophisticated analytical techniques. Labeling is required in the United States to inform the customers that the food has been irradiated and cleansed of any unwanted contamination. The international symbol for irradiated foods is the radura, a green, stylized flower symbol shown in figure 32. Labeling must also include the words "Treated by Irradiation" or "Treated with Radiation." However, since the US government lacks jurisdiction over establishments selling prepared foods, irradiated foods could be

Figure 32. Radura symbol that labels irradiated food.

served without such labeling. Furthermore, irradiated ingredients need not be identified as such on labels of prepared and processed foods. An example would include seasoning spices.

It should be noted that the goal of food irradiation is not to totally eliminate biological contamination, but rather to reduce it to about 0.001 percent of its original value (i.e., reduce the contaminants by about five orders of magnitude). It is important that a small residue of pathogenic microorganisms remains in a healthy body in order to keep our immune system functioning. Without an active immune system, we would be forced to live in a completely sterile environment. This would leave our bodies susceptible to attacks from even minute amounts of foreign substances—and we would have no defense systems left to ward off severe illness or death.

What Are the Benefits of Food Irradiation?

The process of food irradiation inhibits sprouting in vegetables and kills harmful bacteria that cause food-borne diseases such as *Salmonella* in poultry and some seafood, *Vibrio cholera* in fish, or parasites such as *Trichinela spiralis* in pork. It also is very effective in killing *E. coli* in beef and poultry. As such, this process offers an immense improvement in the shelf life and the safety of foods—especially in situations or countries where there is little or no refrigeration. By enabling long-term food storage without decay, impressive contributions can be made to the world's ability to transport food long distances. This substantially enhances world commerce, but it also greatly enhances possibilities for providing food for nations in essentially any part of the world that may have a particular need.*

The ability of food irradiation to rid contaminated food of unwanted pathogens and bacteria is of paramount importance to society. Whereas some citizens continue to labor under the impression that modern medicine can cure any malady, the fact is that our bodies are gradually becoming immune to some of the standard antibiotics. Our immune system ordinarily does a marvelous job in fighting off foreign matter such as bacteria, but it also can become efficient in nullifying the effectiveness of the very medication that we take to fight diseases. There are numerous instances in which physicians have to increase the dosage of antibiotics as a function of time to fight the disease or switch to even-more-powerful medications.

Many epidemiologists are becoming concerned that we need to be giving more attention to improving the safety of our food, rather than attempting to deal with the sickness that follows from eating contaminated food. We need to prevent illness from happening in the first place. Anyone who has suffered through food poisoning will quickly tell you that it is not an experience he or she would like to repeat! The five thousand annual deaths from food poisoning in the United States further attest to the gravity of the situation.

One of the prime advantages of food irradiation is that it sterilizes food without altering significantly the sensorial characteristics of the foods, such as taste or texture. The older methods of food processing, which rely on temperature extremes (heating or freezing), extreme drying or salting, or chemical treating, often change the nature and taste of the food that is treated.

How Carefully Has Food Irradiation Been Studied?

Although food irradiation is relatively new as a commercial process, this technique has been studied more thoroughly than any other food-preservation technology. Over fifty years of research has shown conclusively that there are no adverse effects from the consumption of irradiated food. Extensive worldwide research has unequivocally demonstrated that food subjected to established levels of ionizing radiation presents no toxicological hazard and introduces no special nutritional or microbiological problems.

The following list[13] provides a brief history of the development and testing of the food-irradiation process:

---

* Whereas it would intuitively seem obvious that avoiding food spoilage would always be an attribute, the complex economics of the food industry indicate this may not always be the case. Some producers actually benefit from spoilage because it means more produce must be handled (with profits at each step) to satisfy the market.

1953. Research on food irradiation started during the first half of the previous century, but it was greatly accelerated in 1953 as a result of President Eisenhower's "Atoms for Peace" speech delivered to the United Nations on December 8, 1953.

1964. The US Food and Drug Administration (FDA) began approving food irradiation to rid wheat and wheat flour of insects and to control sprouting of potatoes.

1985. The FDA approved irradiation of pork to prevent trichinosis.

1986. The FDA approved irradiation for fruits and vegetables.

1986. The FDA approved the irradiation of spices and seasonings.

1990. The FDA approved the irradiation of poultry to prevent *Salmonella* and other food-borne bacterial pathogens.

1992. The World Health Organization (WHO) endorsed food irradiation, calling it a "perfectly sound food-preservation technology."

1993. The US Department of Agriculture advocated research on irradiating ground beef (a result of the rash of deaths and illnesses caused by *E. coli*–tainted hamburgers from fast-food restaurants in Washington State and elsewhere).

1997. The World Health Organization was joined by the United Nations Food and Agriculture Organization (FAO) and the International Atomic Energy Agency (IAEA) in endorsing the use of food irradiation (with high doses).

1997. The FDA approved the irradiation of red meat, including beef, lamb, and pork to kill *E. coli*.

1999. The US Department of Agriculture approved the irradiation of red meat.

2000. The National Fisheries Institute, along with the Louisiana Department of Agriculture and Forestry, petitioned the FDA to allow for the voluntary use of irradiation processing for crustacean seafood products, including shrimp, crab, lobster, and crawfish.

Is Food Irradiation Safe?

Whereas safety is always a relative term, no other method of food processing has been subjected to as thorough an assessment of safety as that of food irradiation.[14] The tests for wholesomeness and safety of radiation-processed food include checking for microbiological safety, chemical changes, nutritional adequacy, and feeding studies on animals and humans. Such tests have been conducted by the FAO, the WHO, the IAEA, and numerous national scientific groups. None of the studies to date has revealed any special problems in relation to microorganisms.

The chemical differences between radiation-processed foods and nonirradiated foods are too small to be detected easily. The composition of food remains essentially unchanged, although there can be some small losses in certain vitamins. Such losses are, however, also encountered with other methods of food processing—usually to a much larger extent. Furthermore, many times more nutritional loss is experienced by the food during the cooking process. In any event, such losses are usually minor and can be made up from other sources. None of the short—or long-term studies on animals and trials on human volunteers has shown any adverse effects of consumption of radiation-processed foods.

Endorsements of the safety of food irradiation include, among others, the American Medical Association (AMA), the WHO, the FAO, and the IAEA.

Is the Food-Irradiation Process Now Being Used?

As of the year 2000, over forty nations had already approved the use of food irradiation for more than fifty foods or food products. In addition to the United States, several countries of different regions in the world have approved the use of food irradiation, including numerous European Union countries. Irradiated food is now widely accepted for situations in which food sickness could have particularly catastrophic implications. Examples include foods for US astronauts during space mission assignments (where they may be some two hundred thousand miles from the nearest emergency room) and open-heart surgical patients. Groups that could particularly benefit from the large-scale employment of food irradiation include other hospital patients, school pupils, and airline passengers (particularly for long, international flights).

Widespread acceptance of food irradiation by the general public, however, has been slow. Despite clear acceptance by the knowledgeable scientific community, special interest groups have achieved considerable success in using fear tactics to frighten an unsuspecting public and slow its introduction. Even food processors in the past have been reluctant to accept this new technique on a massive scale, partially because their endorsement would indirectly communicate the fact that their current product is less safe. However, massive recalls of hamburger, and the ensuing publicity and litigation, is changing the attitudes toward food irradiation among food processors and the public. Once the conversion to food irradiation becomes widely accepted, the new infrastructure must gear up rapidly to meet the demand. Most analysts believe the cost of food irradiation can be kept to a few cents per pound, which might be offset by the savings in shelf life alone.

J. P. Corrigan, owner of the Carrot Top, Inc., grocery store in Northbrook, Illinois, was perhaps the first grocer in America to experiment with the sale of irradiated foods at the retail level. He explained that once consumers were educated on the pros and cons of food irradiation, they were willing to pay a premium price for the irradiated products.[15] During his presentation to a symposium organized by the IAEA in 1993, he expressed considerable enthusiasm for the commercial future of this processing method. Indeed, his endorsement seems to be gathering a following. Surebeam Corporation,* builders of an accelerator system to irradiate beef with electrons and x-rays, reported the commercial irradiation of fifteen million pounds of ground beef using their equipment in 2002, and they predicted this would rise to three hundred million pounds in 2003.[16] This is still a small fraction of the nine billion pounds of ground beef sold in the United States in 2002, but the inroads being made are impressive. Dr. Mohammad El Baradei, director general of the International Atomic Energy Agency, noted in a speech given on March 12, 2003, in Vienna that there are approximately three hundred food-irradiation installations in operation around the world.[17] This number is almost certain to grow, and quite likely at a fairly aggressive rate.

One of the early pioneers in the commercial food-irradiation field was Sam Whitney, a successful food and transportation executive in southern Florida. In 1990 Whitney built the first commercial facility in the United States dedicated to irradiating food. It was about the same time that the FDA had outlawed the use of ethylene dibromide as a pesticide to control insect larvae in produce such as strawberries and soft fruits, and Whitney knew that radiation was an acceptable substitute. He demonstrated this by irradiating quantities of strawberries for local growers, increasing the shelf life and allowing for longer shipping times. He was more interested, however, in irradiating poultry, since more than 50 percent of all poultry processed and sold contains *Salmonella*.

Most Americans are not aware of this, and poultry suppliers are not concerned since when chicken is cooked, *Salmonella* and other pathogens are destroyed. Nobody eats raw chicken. On the other hand, during the preparation and handling of the uncooked poultry, or in serving undercooked poultry, *Salmonella* can be transferred to cutting boards, plates, and even other foods. So there

---

* I learned during the final drafting of the manuscript that Sunbeam was forced to file for bankruptcy. It had the difficult job of both creating the market as well as building the capacity to serve the market. Unfortunately, it appears that it overcommitted on the capacity side and was not able to generate income fast enough to sustain its obligations. Though the market was growing rapidly, the company's operational costs were more than the current market could support.

was a legitimate need for this process, and Whitney invested his own money in building a company originally called "Vindicator" and later "Food Technology Services" to commercialize this process.

Unfortunately, there was a great deal of resistance to this company by organizations who opposed any and all operations connected with nuclear technology. Whitney was accused of numerous improprieties and false claims, suggesting that he was in it only for his own gain. But Raymond Durante, president of the Food Safeguards Council and an advocate of food irradiation who knew Whitney, attests to his honesty and dedication to provide safe foods for the consumer. "Sam invested his own money, and applied his full energy to demonstrating that food irradiation could provide significant benefits to the public," according to Durante.[18] Because of the adverse publicity caused by these protests, poultry suppliers were not anxious to be associated with this process, and large supplies of chicken were not forthcoming. It is an economic requirement that whatever product is irradiated must be in a large volume to keep the unit cost competitive. Whitney, along with others, was forced to form a new and separate company to market irradiated chicken in order to achieve this objective.

Today, this same process is routinely used in the Netherlands to irradiate milk cartons as part of its efforts to assure the safety of the milk-delivery system. India is now processing twenty tons of spices per day in its Vashi irradiation plant and has a demonstration unit set up in Lasalgaon to irradiate ten tons of potatoes and onions per hour.[19] Indeed, Sam Whitney's pioneering efforts are beginning to become appreciated.

In an effort to sample public opinion on the acceptability of irradiated beef, the National Cattlemen's Beef Association conducted two nationwide surveys in America in 2002. They found in the February survey that 38 percent of Americans would purchase irradiated meat, and by November that number had risen to 48 percent. Ron Eustice, executive director of the Minnesota Beef Council, reported that over four thousand supermarkets in America now carry either fresh irradiated ground beef or frozen products. Major supermarkets such as Safeway, Albertsons, Jewel, Giant Eagle, and Winn-Dixie have signed on to offer irradiated meat at some of their stores. Mr. Eustice went on to speculate that "food irradiation will take its rightful place as the fourth pillar of public health along with the pasteurization of milk, immunization against disease, and chlorination of our water supply—and that will take place in the next decade."[20]

Dr. Elsa Murano, undersecretary for the Food Safety and Inspection Service of the US Department of Agriculture (USDA), reported at the First World Congress on Food Irradiation (May 5–6, 2003, in Chicago) that the 2002 Farm Bill approved by Congress mandates that commodities such as meat and poultry that

Figure 33. Irradiated beef on display in a major US grocery store (photo courtesy of Ron Eustice.)

are treated by any technology approved by the USDA and the FDA to improve food safety must be made available to the National School Lunch Program.[21] She quickly pointed out that food irradiation is included in this mandate.

Given the unprecedented demands that an increasing world population will place on food supplies, there is little question that radiation will become far more visible as a part of our everyday lives through the widespread use of food irradiation. The contributions that radiation has already made to our food supply via crop production, animal improvement, and pest control have been impressive, even though they have not been very visible to the public. I believe it likely that food irradiation will first become established by the fast-food franchises and then become mass marketed by the grocery stores. It will soon be as commonplace as the pasteurization of milk. Who today would consider purchasing anything other than pasteurized milk? The advantages of consuming safe food, along with the additional shelf life afforded by the irradiation process, are almost sure to transform the gigantic global food industry. The only question is how long it will take for us to accept it and then get there.

5. Medicine

Perhaps the most significant success story in harnessing radiation to serve modern humanity over the past half century is in the field of medicine. Both the quality of life and the longevity of citizens throughout the developed world have improved substantially during the twentieth century, largely because of dramatic medical advances.

I have experienced this personally. Anyone who has sat in a doctor's office and heard the word *cancer* can quickly identify with the shock and disbelief that this word immediately conveys.

My ordeal started out in the most routine path imaginable. I was on the faculty at Texas A&M University at the time and routinely took advantage of the marvelous physical fitness facilities available. Aside from the normal stress of any job entailing responsibility, I was in excellent physical condition. Prime of life!

Hence, I did not expect anything but good news from my yearly medical exam at the nearby clinic. Besides, Dr. Mark Higgins was a good friend of mine and an excellent general practitioner. He dutifully took me through the normal drill and then politely asked that I bend over while he felt my prostate gland. I'll admit to a tinge of worry here, because several times in the past the attending physician noted a slight hardening, but subsequent exams always nullified the concerns. Alas, this time my blood sample, taken as a part of my routine exam, revealed an increase in my PSA over the previous year from a level of about 2.8 to 4.1. Whereas the absolute value was not of particular concern, the appreciable increment from past measurements was enough to cause the cautious Dr. Higgins to give me a call and suggest that I undergo a biopsy.

OK. No big deal. I made an appointment with Dr. James Vestal, a very competent urologist, and he proceeded with the biopsy. Those of you among my male readership who have been subjected to this procedure will know that there can be some interesting moments—especially when the samples are taken near the nerves! Bill Cosby would have a field day describing such moments! But Dr. Vestal kept me at ease by using his best bedside manner to assure me that this was just a routine check. Not to worry.

Consequently, I came back to Dr. Vestal's office a week later all prepared to be reassured that there was no problem. How could there possibly be anything wrong with a fine physical specimen like me? "Alan," he said with a somber face, "I regret to tell you that you have prostate cancer." He then went on to try to assure me that I had several options, because we caught it early (Gleason #6). I could consider surgery (radical perineal prostatectomy), external beam radiation, brachytherapy (radioactive seed implants), cryosurgery (freezing of the tissues), and so on. "But," he quickly pointed out, "I know you won't hear a word I'm saying—given your realization that you have heard the *cancer* word—so I suggest you make an appointment to come back in a week with your wife and we'll go through this again." Wise counsel. It is difficult to think clearly when one is in shock.

But to be perfectly honest, I don't think I really felt the deep shock that most people experience. I was well aware of the brachytherapy approach because I had been discussing the beneficial uses of radiation with colleagues and students for quite some time. Granted, it was a much newer treatment than the "gold standard" of surgically removing the prostate gland. But I was of the impression that it was just as good, if not a better method of treatment, since the side effects (incontinence, impotency, and the like) were so much less. However, I also realized that there are many factors that play into such a decision, including the type and maturity of the cancer itself, the age of the patient, the general health of the patient, and so on. Hence, I tried to keep an open mind on this topic.

Since I had a very busy schedule at the university, my wife became my principal research assistant. She scoured the Internet like a lioness in search of food for her young. Although rapidly reaching information overload, she had plenty of ammunition ready to fire at the affable Dr. Vestal when we returned to sort out the options. Knowing that I was a nuclear engineer, Dr. Vestal recognized that I was not afraid of the radiation techniques and quickly assured me that brachytherapy was an excellent option for my particular situation. Given this assurance, I then began the search for the right medical team.

One of the great factors in my favor was that my daughter-in-law, Suzanne Kolb, was working at the Fred Hutchison Cancer Center in Seattle as part of a multiyear global study of prostate cancer. Although not a physician, Suzanne had access to files from prostate cancer patients all over the world and routinely talked to survivors about their experiences. She noted that such patients will often talk more freely as part of an independent research study than they will to their personal doctors—since they often find it difficult to discuss disappointments with the physicians who performed their treatments.

In any event, she quickly assured me that the seed implant (brachytherapy) approach was now widely used and was achieving very good results. Furthermore, I was surprised to hear that much of the pioneering work on this technique was done in Seattle. This immediately piqued my interest because I had moved to Texas from the state of Washington just three years earlier, and it was my intention to return after my four-year agreement with Texas A&M had been completed. With four children and five grandchildren in the Washington State vicinity, the magnet for returning was exceptionally strong.

Given this situation, I then wondered if I should go to Seattle for the treatment, since I would then be nearby for any follow-up work. Fortunately, Suzanne learned that many Texas patients had been going to San Antonio to have Dr. Bradley Prestidge do their seed implants. Dr. Prestidge had worked directly with the key developers of this technique in Seattle earlier in his career, and I quickly learned that many of my colleagues at the university were very satisfied with his treatment approach. Hence, I decided to have the work done there.

I'll never forget a conversation that I had at the hospital with the nurse who was preparing me for the procedure. She asked me how I was feeling. I replied, "Great. If it weren't for those sensitive diagnostic instruments that you people in the medical field use, I wouldn't know that I had a problem—and I would not even be here!"

After a brief discussion with the anesthesiologist, the next thing I remember was waking up in the recovery room every bit as relaxed as encountering a refreshing fall morning. I learned that I was now the proud recipient of 109 seeds of iodine-125, a radioactive material with a sixty-day half-life. Each of these seeds is about the size of a grain of rice (a cylinder about 0.03 inch in diameter and 0.15 inch long). The enclosing cylinder is made of titanium and is so compatible with human flesh that the seeds can stay in the prostate gland for life. They are inserted into the prostate gland by a special instrument. A three-dimensional imaging device is used by the physician to place them exactly where the radiation physicist specifies for optimal coverage of the gland. I was told that my coverage was over 99 percent (i.e., over 99 percent of my prostate gland would be bathed with an appropriate dose of radiation).

The basic idea of this procedure is for the 28 keV x-rays being emitted from

I-125 to kill cancer cells by causing double breaks in the DNA structure of the chromosomes. Cancer cells grow more rapidly than normal cells, and it is during this reproductive cycle that they are most susceptible to radiation. The sixty-day half-life of I-125 was deemed to be appropriate for my type of cancer, since the low Gleason number indicated it was slow growing. For more aggressive cancer, physicians often use palladium-103, which has a half-life of about seventeen days and emits 20 keV x-rays. There is now a newer FDA-approved radioisotope coming into the market, cesium-131, that has an even shorter half-life (about ten days) and it has a bit more energetic x-ray (30 KeV) that may offer additional benefits.[1] The radiation treatment field keeps expanding to provide optimal flexibility to attending physicians.

The bottom line is that I went home the next day (after spending a relaxing evening in a local motel) and jogged a full mile only five days after this procedure. I have not had a "down" day since, and this process was performed over two years ago. My PSA is down and holding steady. I am particularly pleased that incontinence has not been a problem—although I must confess to having more interest in checking out the porcelain when passing by a bathroom.

Figure 34 is a radiograph of my new "gems!" I doubt that many people would be so brash as to share something quite this personal. In fact, Dr. Prestidge was a bit amused when I asked for a copy and told him I planned to publish it. But I feel exceptionally fortunate to be living at a time when such superb medical options are available. Well over two hundred thousand male adults in the United States alone will be told this year that they have prostate cancer. It is the second leading killer among males in the nation (behind heart disease), with over twenty-eight thousand deaths per year as of 2003. The worldwide death rate from prostate cancer is over two hundred thousand per year. Hence, I include this to provide hope to millions of my fellow citizens who will have to face choices of treatment. I'm certainly not suggesting that this will work for everybody. I am not a physician, and I am fully aware that there are times when surgery, hormones, external beam radiation, and other modalities are more appropriate. But I can personally testify to at least one treatment that modern medicine has brought us though a creative use of radiation. I have bet my life on it.

STERILIZING MEDICAL EQUIPMENT

Ever since the advent of germ theory, medical teams have demanded cleanliness as a crucial part of good practice. Knowing that radiation at high levels can kill microorganisms, the medical community quickly recognized the potential for

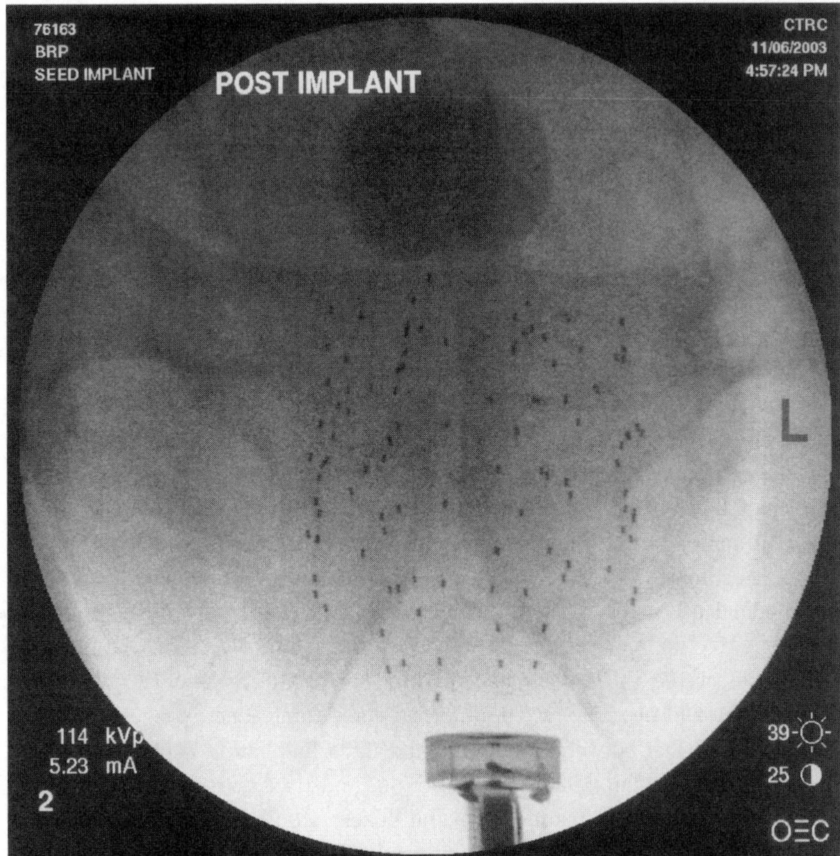

Figure 34. Radiographic image of the author's brachytherapy implant seeds.

employing certain types of radiation (mainly gamma radiation) to sterilize dressings, surgical gloves, bandages, plastic and rubber sheets, syringes, catheters, sutures, heart valves, and a myriad of other devices routinely used during medical procedures. Because radiation is a "cold" process, radiation can be used to sterilize a range of heat-sensitive items such as powders, ointments, and biological preparations such as bone, nerve, skin, and so on used in tissue grafts.

Today, well over half of all sterilized medical equipment used in modern hospitals benefits from radiation treatment. This process is safer and cheaper than most other methods (such as steam) because it can be done after the item is packaged. Hence, the sterile shelf life of the item is then practically infinite—provided the package is not broken open.

NEW DRUG TESTING

It is nearly impossible for today's physicians to deal effectively with severe patient illness without modern drugs. The size of the pharmaceutical industry has recently mushroomed in most developed countries as new drugs are produced to treat previously incurable diseases and anomalies.

But for such treatments to be approved by the designated federal agencies in order to reach the physicians' hands, substantial testing must be done. Mammoth hurdles must be overcome by the drug companies, both to determine how a new product attacks the targeted disease and then to ascertain the potential side effects. Radioisotopes, because of their unique imaging (emission) characteristics, are ideally suited to answer such questions. A radioisotope can be attached to the drug being studied, and the exact location of the drug can be traced through the body by an array of detectors surrounding the patient. Consequently, crucial information can be determined such as material uptake, metabolism, distribution, and elimination of unwanted residues from the body.

It is estimated that over 80 percent of all the new drugs eventually approved for medical use in the United States employ radiation techniques as a crucial component to their success. It should not be surprising, therefore, that radiation techniques played a key role in twelve of the recent fifteen Nobel Prizes awarded in medicine and physiology. The International Atomic Energy Agency (IAEA) has estimated that between a hundred and three hundred radiopharmaceuticals are in routine use throughout the world, and most are commercially available. Appendix A contains a listing of several dozen radioisotopes currently used in the world of nuclear medicine. Appendixes B, C, and D provide comparable listings for industry, environmental protection, and other applications.

DIAGNOSTIC TECHNIQUES

A crucial part of successful medical practice is diagnosing ailments. There are countless examples in every hospital as well as every corner of the globe where an early and exact diagnosis could have prevented tragic results.

It is this element of medicine where radiation techniques have made their most significant contribution to improve health care. The earliest use of radiation for medical diagnosis was employing portable x-ray sources in World War I, where these devices helped field surgeons save many lives. Marie Curie personally delivered such devices to many battlefields during WWI. Today, dental x-rays, chest x-rays, and a plethora of other tests are in routine use in the medical/dental professions.

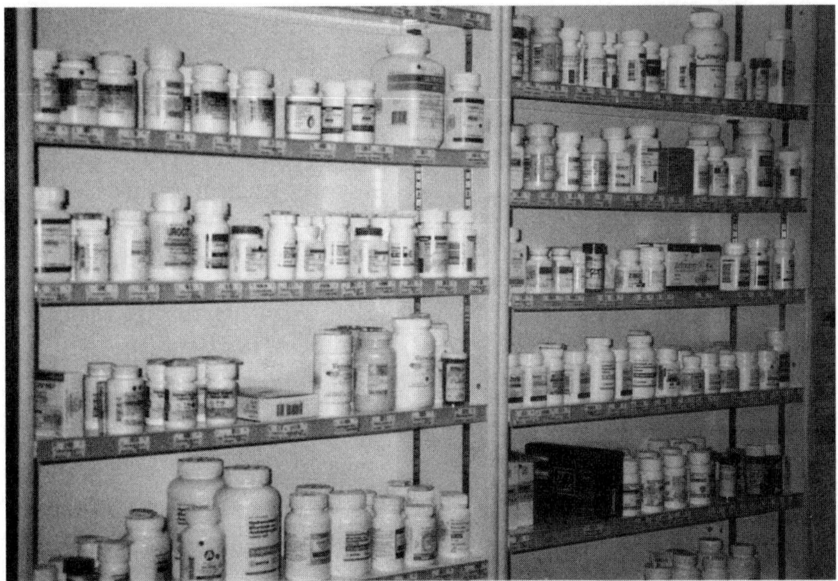

Figure 35. Typical drug store shelf, where over 80 percent of the available drugs benefited from radiation techniques in their preparation.

The basic physics of a successful x-ray is quite simple. Most of us have had an x-ray of a suspicious tooth. As noted earlier, the attending technician puts a film on the back side of the tooth (usually asking the patient to bite down on a "wing" to hold it in place) and then sends a short beam of x-rays through the tooth. The numbers of x-rays reaching the film are directly proportional to the fraction that is not attenuated in and around the tooth. If a cavity exists, more x-rays reach the film than would be the case if the enamel were all intact. Hence, more film exposure takes place at that point, and the dentist knows precisely where to drill.

Approximately two hundred thousand women in the United States are diagnosed with breast cancer every year. Self-examinations and physician examinations to feel physical lumps are certainly two ways to detect a possible cancer. But it is far preferable to discover such malignancies much earlier in their growth cycle in order to maximize treatment options. Hence, mammograms are strongly recommended by most doctors. A mammogram employs the same basic technology as the dentist uses to discover tooth cavities. Each breast is normally x-rayed in two directions. The photographic plates are then analyzed by a radiologist to determine whether there are suspicious spots. Though quite effective in most cases, further advances in this technology have been reported using a dif-

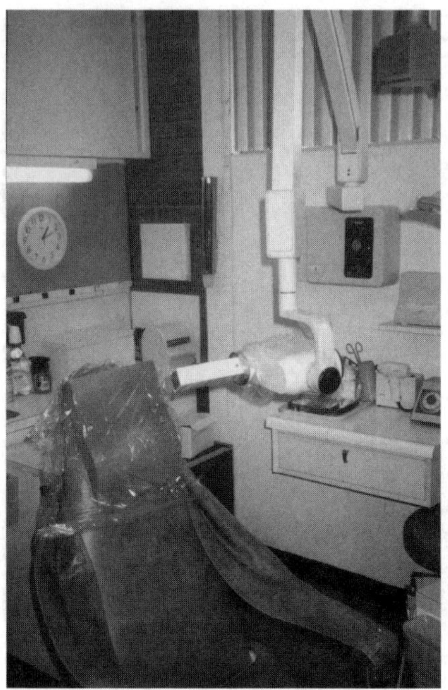

Figure 36. The familiar dental x-ray machine.

fraction-enhanced imaging (DEI) technique to allow higher precision with lower doses of radiation.[2] This effort is indicative of the constant work underway, worldwide, to improve medical diagnostic techniques.

The next degree of sophistication is to employ an x-ray machine that is designed to completely encircle a patient lying on a flat table. This is called a CAT scan (computed axial tomography). By taking x-rays at numerous axial and angular positions, it is possible to merge the images by using computers to form a three-dimensional image from a series of two-dimensional x-rays. This allows physicians to detect potential anomalies deep within the human body without having to resort to surgical procedures to gain direct visual inspection.

But x-rays, useful as they are, provide only a snapshot of a particular piece of the anatomy. The imaging properties of radioisotopes allow modern nuclear medical specialists to measure the performance of some specific physiological or biochemical functions in the body as a function of time. This has enormous implications, all the way from determining nutritional deficiencies to locating and identifying various types of cancer. The latter includes detecting an unknown primary site of cancer in a patient where cancer has metastasized, differentiating benign from malignant lesions, grading the degree of malignancy, understanding the extent of the disease, assessing the response to treatment, and detecting recurrent disease.

To understand how such analyses can be done, let's consider a specific example; namely, the use of a radionuclide in the measurement of blood flow. Prior to the use of radioisotopes, doctors used chemical dyes to measure how much blood the heart could pump. The patient was injected with a known amount of dye, and numerous blood samples were subsequently taken to measure the degree of dilution as a function of time. Needless to say, this resulted in substantial uncertainty. But in 1937 San Francisco neurologist Joseph Hamilton injected a patient with a radiotracer, placed a radiation detector on the heart, and recorded how much of this material passed through the heart as a function of time.[3] He called this process a "radiocardiogram," a procedure still used today, but in a much improved form (different tracers and considerably better detectors).

Substantial progress has been made since that time. It is now possible to conduct dynamic studies of the cardiac cycle to determine the length of time necessary for the heart's chamber to fill and empty, and to precisely examine the movement of the heart's walls. Recent updates of this basic procedure are being used to help emergency room doctors do a better job of ruling out heart attacks in patients with chest pains.[4] This new procedure, called myocardial perfasion imaging, is performed by injecting a radiotracer that provides a picture of the blood flow to the heart muscle, allowing the nuclear medicine physician to determine the risk of the patient having a heart attack.

The four most common detection approaches currently used in modern diagnostic nuclear medicine beyond the CAT scan are single photon emission computed tomography (SPECT), positron emission tomography (PET), nuclear magnetic resonance (NMR) imaging, and radioimmunoassay (RIA). It is interesting to note that the NMR is now commonly referred to as magnetic resonance imaging (MRI), where the word *nuclear* was dropped by the medical community strictly to improve its marketing value. This is sad since its nuclear properties make this technique possible.

SPECT is widely used for routine clinical work because it is relatively inexpensive and utilizes radioisotopes available from nuclear reactors. Technetium-99m*, a very popular 140 keV gamma emitter with a six-hour half-life, is the most popular radioisotope used in this device. It is normally derived from a common nuclear reactor fission product (molybdenum-99), although it is technically possible to make it from other nuclear reactions in a nuclear reactor. Mo-99 has a sixty-six-hour half-life (slightly under three days), and it decays to Tc-99m. The "generator" used to make Tc-99m consists of a lead pot enclosing a glass column that contains Mo-99. When an order for Tc-99m is placed, it is washed out of the lead pot column by a saline solution and prepared for injection into the patient. Because the Tc-99m is "milked" from the parent Mo-99 upon demand, the lead pot is called a "cow." Having been raised on a dairy farm, I have always been fond of this analogy. After about one or two weeks of use, the generator (or "cow") is returned to be supplied with a new batch of Mo-99.

---

* The *m* in Tc-99m refers to a metastable state of this isotope. By emitting a gamma ray, the high energy state of this isotope is lowered to its ground state, and it becomes the nearly stable isotope Tc-99. One of the fascinating aspects of atomic physics is that Tc-99 is still slightly radioactive, but its half-life is over two hundred thousand years.

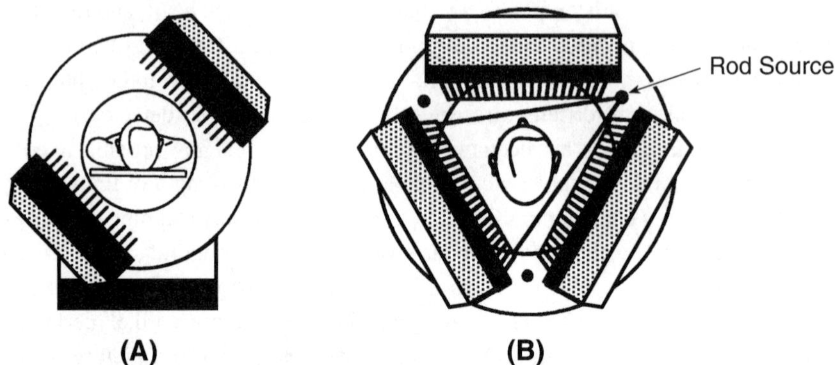

Figure 37. SPECT imagining sketches.[5] (A) Dual-head rotating gamma camera system; (B) Triangular system with rod source for calibration.

Figure 37 illustrates how a SPECT system might work. The left illustration (A) depicts a dual-head rotating gamma camera system, wherein the patient stays in a fixed position and the two cameras (detector systems) rotate around the patient, picking up the gamma rays emitted by the Tc-99m circulating in the patient's body. By the clever use of microprocessors, the data collected by the cameras can be sorted out and the location of the Tc-99m radioisotope can be displayed in three dimensions and also as a function of time.

If bone cancer exists, the specially prepared chemical carrier (methyelene diphosphorate, or MDP) to which the Tc-99m is attached will concentrate abnormally at the sites of the bone reactions to the tumors, and the "tell-tale" increased bone uptake of Tc-99m—MDP—at those sites clearly reveals the problem. It was the lack of such "hot spots" that allowed the attending physician to assure my mother, as mentioned in the opening chapter, that she did not have bone cancer. If the physician is looking for other types of abnormalities, a different chemical carrier is used (one that has a propensity for accumulating at the suspicious sites). Nuclear medicine researchers have developed a large number of such radiotracers that allow visualization and measurement of the functions of almost any organ in the body. Since most diseases produce abnormal function before there are abnormalities of anatomy or structure, these functional nuclear medicine techniques usually allow early diagnosis, when treatment is most effective.

The right illustration (B) of figure 37 depicts a more sensitive arrangement of the SPECT configuration. This system, which employs three camera heads, is particularly attractive for brain imaging because it employs a set of radioactive rod sources to help calibrate the detectors in accounting for attenuation variations during the measurements. These improvements also help in imaging other parts of the body. Several other arrangements are employed to help sharpen the images with enhanced speeds, but all SPECT systems work on the fundamental principle of attaching a radioisotope to a specified chemical carrier that travels through the body and is traced by radiation detection with a special camera system. Whereas Tc-99m is by far the most popular radioisotope used for such purposes, many other radioisotopes made from both reactors and cyclotrons are currently in use. These include iodine-131, iodine-123, gallium-67, thallium-201, and indium-111.

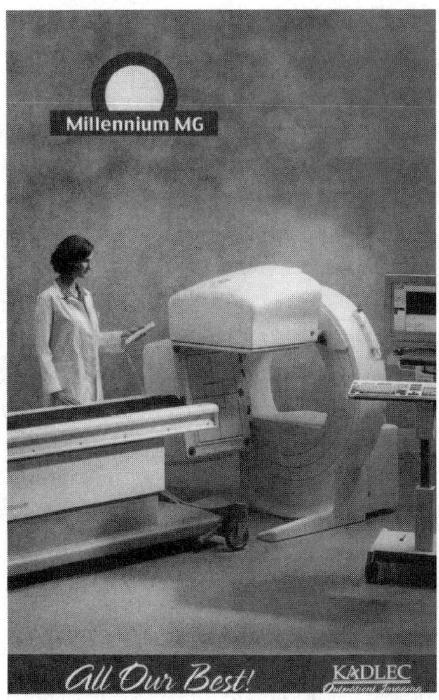

Figure 38. A modern SPECT machine. (Photo courtesy of Kadlec Hospital, Richland, WA.)

It is estimated that Tc-99m is used in over thirty thousand procedures every day in the United States alone. This amounts to over ten million procedures every year! The late professor Glenn Seaborg, Nobel Laureate who directly served ten US presidents in numerous capacities (including chairman of the Atomic Energy Commission and chancellor of the University of California, Berkeley), is credited with the codiscovery and development of this precious radioisotope. He is quoted as remarking that if he had done nothing else in his entire career, the discovery of this radioisotope would have provided more than ample satisfaction for a productive life. Nuclear diagnostic techniques, based largely on this technique, are so pervasive that one out of every three patients who enter a US medical center today directly benefits from nuclear medicine. Fortunately, Dr. Seaborg lived to witness the incredible impact of his work.

PET devices are based on the detection of a pair of photons emitted from positron annihilation. Very shortly after a positron (an electron with a positive charge) is emitted from a radioactive substance such as flourine-18, it collides with an electron causing a literal annihilation of the two particles. The mass of the two particles is transformed into pure energy (according to Einstein's famous $E = mc^2$ law, where E is energy, m is mass, and c is the speed of light) producing two gamma rays of exactly 511 keV each that move apart at light speed in precisely opposite directions. By surrounding the treated patient with special detectors, the location of the radioisotope can be pinpointed by determining counts recorded at exactly the same time (coincidence counting) on opposite sides of the patient.

PET systems tend to be more expensive than SPECT systems, partly because of the sophistication of the counting system and partly because the radioisotopes that emit positrons typically have a very short half-life (minutes or hours). Hence, they must be produced either on-site or in close proximity by accelerators (usually cyclotrons) and administered to the patient with the proper chemical carrier very quickly. But PET machines are becoming increasingly popular because they are capable of visualizing F-18—FDG (flurodeoxyglucose) uptake throughout the body, which allows high-quality three-dimensional images of glucose-avid cancers with greater sensitivity than any other imaging technique. These systems are now being coupled with fast CT scanners to provide both the functional image (PET) and a detailed anatomical map (CT). As an added feature, the deoxyglucose can traverse the blood/brain barrier and detect abnormalities in the brain. Radioisotopes often used in such devices, in addition to F-18, include carbon-11, nitrogen-13, oxygen-15, and rubidium-82.

Magnetic resonance imaging (MRI) devices are also used for diagnostic purposes, but they do not use radioisotopes (no ionizing radiation). Rather, they rely on the phenomenon that a strong magnetic field applied to the organ under investigation will cause the nuclei in that organ to preferentially align their nuclear spins parallel to the magnetic field. When photons of a certain frequency from an external source are transmitted into this field, the nuclei absorb this energy, and this causes the nuclear spins to change their orientation and oppose the magnetic field. When the source of external photons is turned off, many of the nuclei will revert to their lower energy state by emitting photons at characteristic resonance frequencies, thereby revealing important information about the organ under investigation. Because the protons in biologically abundant hydrogen are particularly sensitive to shifts in magnetic gradients, it is possible to determine density variations of bone marrow and tissue, thereby obtaining excellent cross-sectional images of the body and its associated organs.

It is even becoming possible to use this approach to determine where different mental tasks are performed in the living brain. This is done by tracking the extra blood flow to the brain's active regions. Because of these attributes, this technique has become a very popular imaging device used for diagnosis in the radiology department of almost all large hospitals. It should be clear that this technique fundamentally relies on the properties of nuclear magnetic resonance and, as such, was originally called the nuclear magnetic resonance imaging (NMRI) technique. As noted earlier, this designation was euphemistically changed by the medical marketing community to MRI, that is, taking the word *nuclear* out, purely for better patient acceptance.

Another diagnostic tool that has enjoyed considerable use is called radioimmunoassay (RIA). This process, discovered in 1960, was developed by Dr. Roslyn Yalow, who was later awarded a Nobel Prize in Medicine for her work. It is based on a chemical procedure using radioisotopes to determine the concentration of biological materials in the human body very accurately—in parts per billion and even better. The human body has a very sophisticated immune system. Whenever a foreign body attacks a cell, a protective substance (antibody) is produced to encounter the foreign (antigen) attacker, and the antibody and antigen react. This is the normal process that takes place during vaccinations, immunizations, and skin tests for allergies.

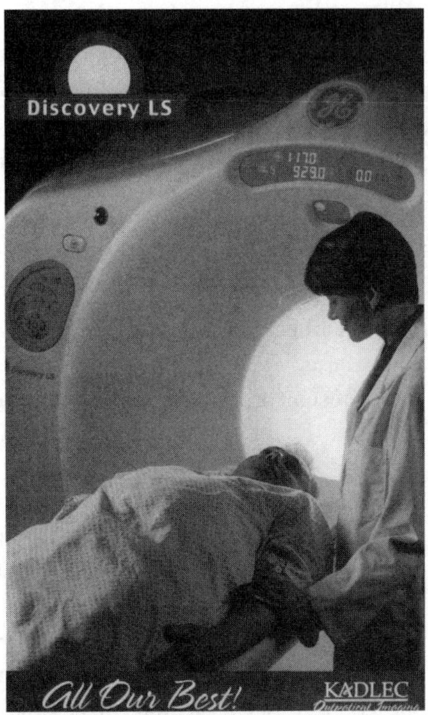

Figure 39. A modern PET machine. (Photo courtesy of Kadlec Hospital, Richland, WA.)

The object of the radioimmunoassay approach is to measure the type and amount of an antigen (unidentified and unwanted attacker) present in a patient. By employing the RIA technique, these results can be determined by analyzing a sample of blood extracted from a patient. By inserting a known antigen with an attached radioactive material into the sample, a competition will then result between the known antigen (the one with the radioactive signature) and the unknown antigen, in that they both will react with the antibody. A chemical or mechanical separation is then performed, and the radioactivity in the products is compared with those in a standard reaction. This technique has been and is continually being used to detect a wide variety of otherwise unknown substances, including hormones, enzymes, the hepatitis virus, and some drugs. It is also useful in identifying nutritional deficiencies. A particular attribute of this procedure is that the patient never comes in contact with the radioisotopes used in the diagnostic tests, since the procedures are all performed on the patient's blood that has been taken into a laboratory. Whereas the RIA technique is still the most precise method for measuring antigens and nutritional deficiencies, other nonradiation methods are now enjoying increasing popularity.

Nuclear diagnostics are now routinely employed throughout the developed world to determine anomalies in the heart, brain, kidneys, lungs, liver, breasts, and thyroid glands. Bone and joint disorders, along with spinal disorders, also benefit directly from this routine use of radioisotopes.

Some of the more fascinating aspects of nuclear medicine help explain how our brain functions. In the 1970s Louis Sokoloff and his colleagues demonstrated that fluorine-18 deoxyglucose (F-18 FDG, a positron emitter), which we discussed earlier, could be employed to measure the use of glucose in specific parts of the brain.[6] This was an important breakthrough because very active brain tissues consume more glucose than average brain tissues. Since malignant tissues grow quite rapidly, their existence and location can be detected through this process. Taking this to the next step, it is now possible to examine the role of chemical recognition sites in living human brains. This allows specialists to determine how humans encode sensory input, when and how we store it, and how this is all put together with other encoded information to produce actions. In other words, we now appear to be on the threshold of determining how our brain accepts input and makes decisions on appropriate responses to this input.

Being able to obtain real-time biological activity in sensitive areas such as the brain allows a plethora of new possibilities for diagnosing abnormalities. Brain scans conducted by either SPECT or PET devices can now detect altered metabolic activity in both temporal lobes of the brain to reveal the onset of Alzheimer's disease.

Another dehabilitating disease that strikes many elderly citizens (including a few young people such as the popular actor Michael J. Fox) is Parkinson's disease. It has been known for quite some time that dopamine concentrations are low in the brains of patients with Parkinson's disease. At an even earlier time, there was evidence that an extract of an Indian herb, *Rauwolfia serpentine*, was effective in reducing the concentrations of dopamine in various organs, such as the brain, heart, blood vessels, intestines, and adrenal medulla.[7] It was only when the active ingredient of this herb was labeled with carbon-11, a radioactive tracer, that the physiological processes causing such concentration changes were understood. The drugs were found to act by blocking the binding of dopamine (a neurotransmitter) to the dopamine receptor. Based upon this finding, a precursor of dopamine, the amino acid L-DOPA, was developed to increase the concentration of dopamine in the brain and relieve many of the symptoms of Parkinson's disease.

Studies on Parkinson's disease have also been under way in France using the microbeam synchrotron x-ray fluorescence technique at the European Synchrotron Radiation Facility to determine the role of trace metals in Parkinson's disease and amyotrophic lateral sclerosis (ALS).[8] By determining the levels of metals such as copper, zinc, iron, and selenium in the degraded neurological tissues, there is hope that deficiencies can be identified and appropriately supplemented.

In addition to the accuracy in determining medical abnormalities that nuclear diagnostics provides to the physician, a great advantage to the patient is that he feels no discomfort during the test, and after a short time, he is left with no traces that the test was ever done. The radioisotopes simply decay and disappear completely. The noninvasive nature of this technology, together with the ability to observe an organ functioning from outside the body, makes nuclear diagnostics a very powerful tool. Patients are spared the trauma of exploratory surgery and possible infections.

Therapeutic Approaches

Until relatively recently, ionizing radiation has been used more for diagnostic techniques than for actually curing diseases. However, some therapeutic work has been underway for several decades. One of the first therapeutic uses of radioisotopes was employing iodine-131 to cure thyroid cancer, an approach introduced in the early 1940s. Since the thyroid gland has a special affinity for iodine, it is a relatively simple and straightforward matter to have a patient drink a carefully determined amount of I-131 in a palatable form of solution. The I-131 then preferentially lodges in the thyroid gland and the beta-emitting properties of this radioisotope subsequently target and destroy the thyroid malignancy. Since I-131 has a half-life of eight days, it does its job and then effectively disappears within a few weeks.

Today, radiation is used quite frequently in the treatment of other cancers. Surgery, chemotherapy, and radiation (often used in combination) constitute the principal venues of cancer treatment.

Most of the current procedures utilizing radiation to kill cancer cells in humans are based on delivering the radiation to the patient externally. This is called *teletherapy*. Accelerators are used to deliver either protons to the target (such as the system used for external-beam prostate treatment at the Loma Linda facility in California) or more often electrons, which are normally directed onto a target that secondarily produces x-rays. Such treatments have achieved very effective results for many types of cancer. However, it is nearly impossible to keep the radiation from killing or impairing some healthy tissue in the immediate vicinity—especially if the beam must pass through healthy tissue to reach the malignancy.

The three principal approaches underway to prevent radiation therapy from injuring healthy cells are conformal radiotherapy, creating radioisotopes at the site of the malignancy, and developing a method to deliver appropriate radioisotopes directly to the cancerous tissue.

Conformal radiotherapy is a procedure by which the incident beam of x-rays or protons is moved around the patient in a precise manner so that the beam remains focused on the tumor, but the penetrating radiation beam does not remain on any of the healthy cells for a very long time. This is sometimes referred to as the "gamma knife."

An example of the second approach is called boron-neutron capture therapy (BNCT). Boron is injected into the patient as part of a special chemical carrier such that it preferentially concentrates at the tumor site. A neutron beam is then focused on the boron, producing alpha particles that destroy the malignant cells only in the immediate vicinity of the concentrated boron. Since alpha particles are stopped at a very short distance from their point of origin (approximately one human cell), the intense radiation damage is very localized. Though this approach has been under experimentation for over two decades, it is only recently that the technique has become sufficiently sophisticated to begin human trials. Such trials are now underway at the Massachusetts Institute of Technology, among others places, where the 5 MW research reactor in the Department of Nuclear Engineering has been retrofitted with upgrades to perform treatments for glioblastoma multiforme (GBM) brain cancer and melanoma.

Some damage may be done to healthy cells through which the neutrons must pass to reach the malignancy, but special "beam tailoring" can be done to minimize this deficiency. Some clinicians prefer to open the skull for brain tumor irradiation to allow the neutron beam clear access to the target (assuming the tumor or the cavity remaining after surgery is located where such access is possible). Also, in some cases it may be possible to remove the affected organ and irradiate it externally, thus avoiding all secondary damage. A "first-of-a-kind" example of this approach was the removal of the liver of a forty-eight-year-old man who was diagnosed with fourteen tumors in his liver. His liver was surgically removed, irradiated using the BNCT technique, and then reattached in a twenty-one-hour operation. One year later, the patient was doing well and had a normally functioning liver showing no signs of tumors.[9]

An example of the third approach is cell-directed radiation therapy. In order to attain the localized damage desired, either beta or alpha emitters are needed. For solid tumors, one method of getting the radioisotope to the target is direct injection, assuming the tumor is accessible. The most straightforward way to treat a localized malignant tumor is to place a radioactive source directly next to it. Iridium-192 is often used for such purposes. It is produced in wire form and introduced through a catheter to the target area. After administering the correct dose, the implant wire is removed and placed in shielded storage.

Another example of this approach is brachytherapy.* The term "brachytherapy" is derived from two ancient Greek words: *brachy* (meaning short distance) and *therapy* (meaning treatment). As discussed earlier, one well-founded application of this technique is treating prostate cancer. This is accomplished by encapsulating a small amount of a radionuclide such as I-125, Pd-103, or Cs-131 within a titanium capsule about the size of a grain of rice. These "seeds" are then placed directly into the prostate gland where they remain for life, as noted in my personal story. Tests are currently underway to use this technique for curing cervical cancer, endometrial cancer, and coronary heart disease.

Substantial progress is also being made using this technique to treat breast cancer.[10] Whereas the recommended treatment for breast cancer a decade or so ago was mastectomy, evidence appears to be mounting that lumpectomy (removal of only the localized malignancy) followed by radiation is equally effective (and far less disfiguring). The radiation treatment is needed to kill any stray cancer cells remaining nearby after the surgery. Relatively low doses of external beam radiation are normally prescribed to reduce unwanted side effects of radiation impinging on healthy cells. However, this means that the radiation treatment must be extended over several weeks to assure the destruction of stray cancer cells.

A possible new route is to use the brachytherapy technique. Immediately following the removal of a solid malignant tumor in the breast, a balloon catheter is inserted into the cavity previously occupied by the tumor. A radioactive material such as I-131 is then loaded into the catheter for a period of a few days to allow the emitted beta particles to kill any cancer cells left behind. Over twenty-two hundred patients at some 250 health care centers have been treated by one such method, wherein radioactive seeds have been placed inside the breast for about five days and then removed. Although sufficient time has not yet elapsed to determine whether there are any negative long-term side effects, these early results appear very promising.[11]

The brachytherapy approach practiced today utilizes primarily x-rays for killing cancer cells, although beta irradiation is employed in some instances. Unfortunately, there are some types of cancer that are quite resistant to alpha, beta, or photon sources. Such cancers include brain tumors, melanoma, sarcoma, locally advanced breast cancer, cervical cancer, and cancer of the head, neck, mouth, and even certain types of prostate cancer. Neutrons appear more effective

* In the 1920s in France, this technique was called Curie Therapy. It was used in the treatment of uterine cancer because it is relatively easy to place the radioactive emitter via a needle in an optimal position.

in treating such malignancies. Today, the only technique to administer neutrons internally is the BNCT approach described earlier. But scientists at Oak Ridge National Laboratory have been working with the private sector to produce miniature sources of californium-252, a very heavy man-made neutron-emitting isotope normally used for large-scale industrial diagnostics purposes (see, by way of comparison, chapter 7). It now appears possible to reduce the diameter of the source Cf-252 material to the point where it can be inserted into seeds the size now commonly used for prostate cancer treatment. Whereas this process has yet to begin clinical trials, there are high expectations—especially for treating glioblastoma multiforme, the most common primary brain tumor.[12]

It may be instructive at this point to note that very extensive and carefully conducted testing must be done on humans before any new medical approach can be licensed by the US Federal Drug Administration (FDA). Phase I tests can be conducted only on patients who have failed all licensed modalities and would otherwise die. If successful in the Phase I trials, the new modality can be tested with increasingly larger groups of patients and at an earlier stage of disease development in Phase II and Phase III. This process often takes several years to complete. Needless to say, any new treatment that is able to deliver a high remission rate in Phase I trials holds immense promise—since only patients who are severely ill (and have failed all other treatments) are allowed to participate. We might wonder why the FDA requires such extensive further testing under these favorable conditions (given the expectation that much healthier patients would respond with even greater remission). But the FDA wants to have a high degree of confidence that secondary, long-term, potentially dangerous side effects do not nullify the advantages of the new treatment.

Returning to our review of cell-directed radiation therapy, an effective approach is to find a chemical that has a special affinity for the malignancy, and then attach the radioisotope to this special carrier. This is called the monoclonal antibody (or "smart bullet") approach. Such a method is particularly suited for treating malignancies that are not confined to a particular spot. Leukemia and Non-Hodgkin's diseases are examples. Recent work employing the "smart bullet" approach has revealed some very impressive results. End-stage Non-Hodgkin's disease has been treated with yttrium-90, with a remission rate of over 80 percent (for patients who have failed all other known treatments).[13] This means that 80 percent of these treated patients have lived for at least another five years without a recurrence of their afflicting disease. Zevalin is the trade name for one of the new drugs based on yttrium-90. Patients with advanced stages of B-cell lymphomas, treated with a similar monoclonal antibody labeled with iodine-131, have a demonstrated survival rate of over 90 percent. Bexxar is the trade name for this drug, recently approved by the FDA.

A variation of this approach is called radioimmunotherapy (RIT). This is a procedure for treating minimum residual disease (e.g., the likelihood that cancer cells may have been left behind after a lumpectomy). Beta sources such as copper-67 and yttrium-90 have been used for these purposes. Since this technique is still relatively new, it is not yet clear whether treatment is best administered using only the RIT technique or whether it should be accomplished as part of a multiple-modality approach (i.e., in combination with external beam radiation or other techniques). Dr. Sally DeNardo, professor of internal medicine (hematology, oncology) and radiology (nuclear medicine) at the University of California, Davis, has pioneered much of this work and reports that RIT agents are in Phase III testing for lymphoma and in Phase I or II for breast cancer.[14]

Recent trials of cell-directed radioisotopes using an alpha emitter (bismuth-213) have shown remarkable results in treating leukemia.[15] Alpha-particle irradiation appears to hold particular appeal for cell-directed radiation therapy because the "kill" range of the alpha particles is so localized. Because of the growing interest in using alpha emitters for this approach, this monoclonal antibody technique is sometimes referred to as targeted alpha therapy (TAT).

It is interesting to note that one of the most valuable radioisotopes necessary to fuel the future of "smart bullet" development using alpha emitters is radium-226. This is the radioisotope that Marie Curie discovered in 1898 (over a hundred years ago). Radium-226 serves as the target material that can be irradiated in a nuclear reactor to produce several of the alpha emitters under current study for irradiation of nonsolid malignancies. Substances such as actinium-227 (which decays to radium-223), thorium-228 (which ultimately decays to bismuth-212), and thorium-229 (which decays to actinium-225 and ultimately to bismuth-213) are all derived from the irradiation of radium-226. One of the fascinating features of the radioisotope actinium-225 is that when it decays, yielding an alpha particle, several daughter products are formed—each subsequently decaying with additional alpha particles. Dr. David Scheinberg of the Memorial Sloan-Kettering Cancer Center in New York has experimented with this approach using mice that were purposely injected with human cancer cells. He then set up two control groups, one treated with Ac-225 and the other with no subsequent treatment at all. He noted that the mice having no subsequent treatment lived an average of forty-three days before dying of cancer. Mice receiving the Ac-225 lived up to three hundred days, with those receiving the highest radiation dose living the longest. This success spurred considerable interest in the medical community, and human tests are now underway.[16]

Several areas of treating specific abnormalities are developing on almost a constant basis. A typical example is treating our circulatory system. Most people are aware of a procedure called angioplasty (that of inserting a "balloon" into a clogged artery and passing it through in a "Roto-Rooter" manner to unclog it). Whereas this procedure has a high success rate, and has prevented a plethora of heart attacks, there are several cases (upward of 40 percent) where the arteries slowly become reblocked. Several years ago, it was discovered that lining the "balloon" with rhenium-186 made a huge impact in preventing reclosure of the arteries. Whereas physicians are yet to be in complete agreement regarding the precise explanation for this success, it appears that the soft beta particles emitted from this radioisotope limit the development of scar tissues on the artery walls, thereby preventing the vessels walls from closing up again. National news was made a few years ago when it was reported that Vice President Dick Cheney would likely be recommended this treatment if his previous treatments proved to be ineffective.

In another example relating to blood, situations often arise in which blood transfusions are desired for patients who are immune-deficient, such as cancer patients or patients receiving organ transplants. One of the dangers of such transfusions is the transmission of lymphocytes, a malady that can cause "transfusion-induced graft versus host disease" (T-GVHD). In some instances the body will not accept blood from another person, and complications resulting from this problem can be fatal. To prevent such concerns, donor blood is often irradiated with a cobalt-60 or cesium-137 gamma source to inhibit lymphocyte proliferation. Even for cases in which the recipient has a sufficient immune system, if the donor is a close relative, the lymphocytes can get past the immune system and go undetected like a Trojan horse. Hence, treatment of the blood via radiation is generally considered by physicians to be a sure way to prevent such complications.

An example of another specialty area is the treatment of arteriovenous malformation (AVM). This condition is a malformation of blood vessels characterized by a mass of unwanted arteries in the brain. A special mixture containing a radioactive powder is injected into the artery, causing an arterial occlusion, thereby stopping the blood flow into the unwanted vessels. This is but one example of numerous applications of radioactivity to somewhat unique conditions.

Although many of the above results are still in relatively early trial stages, the potential for success is enormous. Given that cancer and heart disease remain of paramount concern in most areas of the world, and that cancer is the most prevalent childhood disease in the Western world, the incentive for further use of radiation in the field of medicine remains huge.

Finally, there are certain types of cancer that are not yet treatable but cause excruciating pain. Bone cancer is perhaps the best example at present. I vividly recall listening to a talk show several years ago in Washington, DC, while driving down the George Washington Memorial Highway. An elderly gentleman was explaining to his talk-show host that he had bone cancer and that the pain was becoming unbearable—so much so that he had decided to commit suicide. Despite the plea from his host to confide his anguish to his children, he steadfastly stated that he did not want his family to know. Regrettably, sometimes it is easier to vent to a stranger than to one's loved ones. Having reached my destination, I felt a throbbing in the pit of my stomach since I had become vaguely aware that certain radioactive substances were nearing approval from the FDA to be used for bone cancer palliation (pain relief). I subsequently learned that both strontium-89 and samarium-153 do have approval from the FDA for such use. Unfortunately, they will not cure the disease, but they will allow many patients to enjoy the last months or years of life without the constant pain associated with this dehabilitating and exceptionally painful disease.

The 2003 calendar factual narrative, issued jointly by the Nuclear Medicine Industry Association of North America, the Society of Nuclear Medicine, the American College of Nuclear Physicians, the Academy of Molecular Imaging, and the American Society of Nuclear Cardiology, reported that on the average, one out of every three hospital inpatients undergoes one nuclear medicine scan during his or her stay. Over a hundred nuclear medicine diagnostic and therapeutic procedures are available in the United States for routine use today—and many more are in the clinical testing stages.

We can all be very grateful for the contributions that radiation has made to the field of medicine. What a great time to be alive!

6. Electricity

Having grown up on a small farm, I often remember the saying "You don't miss the water until the well goes dry." I learned its wisdom on several occasions each summer—with warnings from Dad to not even consider watering the lawn or washing the car. Precious water was needed for washing the milking utensils and cleaning the milking parlor. After all, that was our livelihood.

The same can be said for electricity. Even back in the midfifties, when I was called out of bed at 5:00 AM to help with the morning milking operations, a loss of power meant a near catastrophe. I always dreaded a large windstorm or heavy snowfall, because that often would result in a tree across the road—severing the power line that snaked up our small Bunker Creek road. Not only did the lights go out, but the milking machines went dead. That meant hand-milking some two to three dozen cows among the mixture of our Guernsey and Holstein cattle. Not much fun—especially in the dead of winter.

And this dependency upon electricity has grown ever so much more in the past half century and beyond. Today, over one-third of the total energy consumed in the United States comes in the form of electricity. It powers almost everything that we take for granted in our everyday lives. We start our day with "lights on," and many of us wake up to a soothing hot-water shower, followed by use of a hair dryer to allow us to enter the outside world, and then sit down to some scrambled eggs or hot oatmeal for breakfast—along with a steaming cup of coffee. Some of those hot-water tanks and kitchen stoves are heated by natural gas, but an even higher percentage of them derive their energy from electrical heating coils. The hair dryer, coffee pot, and toaster must be plugged into the electrical socket. Computers, printers, fax machines, plus a myriad of motors and other devices commonplace in modern commerce also all rely upon electricity.

All we have to do to appreciate the conveniences that electricity brings us is to spend an hour or so trying to survive a blackout. Simply put, we are hooked on electricity and will become even more so in the future.

Given this, let's now step back and take a brief historical peek at how radiation became a key player in the production of electricity.

HISTORY OF NUCLEAR POWER

In one of the most dramatic moments of his presidency, Dwight Eisenhower mounted the podium at the United Nations in New York City on December 8, 1953, to deliver his famous Atoms for Peace address.

To be sure, one of Eisenhower's principal concerns was to strive for ways to limit the proliferation of nuclear weaponry, since the United States at that time had a clear international lead in the design and production of such weapons—based upon the splitting of the atom. In many ways, it is unfortunate that the world was first introduced to the power of nuclear fission via the atomic bombs dropped over Japan to end World War II. I, along with many of my colleagues, have often wondered how atomic power would currently be viewed by the general populace if nuclear fission had first been harnessed to light schoolrooms, rather than to destroy a major city. Nonetheless, world conditions of the 1940s naturally led to military, rather than civilian, priorities.

Given this historical background, President Eisenhower appealed to the world community to use the awesome power of the atom for peaceful, rather than military, purposes. It was within that spirit that the United States, along with several other nations, began an earnest campaign to exploit nuclear fission to power lightbulbs, rather than war devices.

Figure 40 illustrates the buildup of nuclear power plants in the United States, which today has resulted in 103 operating commercial plants delivering approximately 20 percent of all the electricity currently consumed in our country (serving one out of every five homes and businesses). In addition to the total installed capacity (now approximately 100,000 megawatts) another twenty-eight nuclear plants were started up at some point during this four-decade buildup, but they were shut down for a variety of reasons. Many of these were early prototypes that were not expected to stay on-line for more than a few years, but others admittedly suffered from immature quality controls and inexperienced manage-

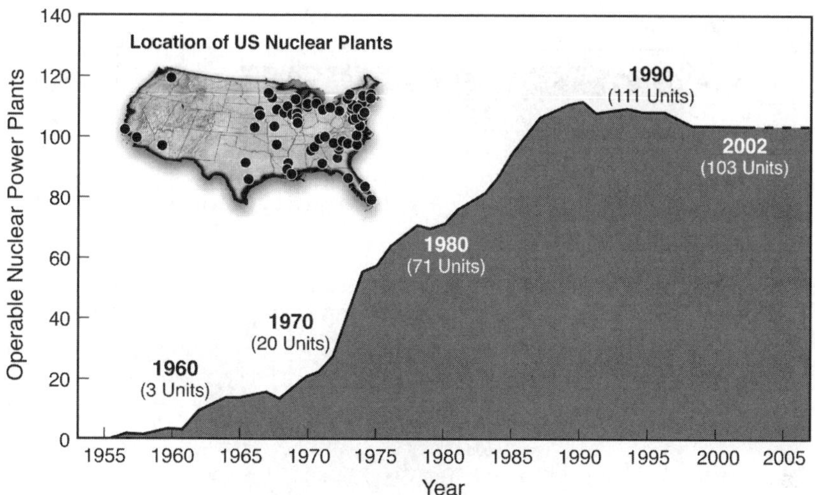

Figure 40. Commercial nuclear power plant buildup in the United States. (Data courtesy of the Nuclear Energy Institute.)

ment. Several others ran for several decades but have now been shut down and decommissioned. Many orders for new plants, originally placed in the 1970 era, were cancelled because of what was later revealed to be overly optimistic growth projections for electricity demand.

To provide a rough measure of the impact of nuclear power, a 1,000 MW plant generates enough electricity to serve a population of about five hundred thousand people (i.e., a city about the size of Austin, Texas). Hence, the 100,000 MW of nuclear capacity now operating in the United States provides all of the electrical needs of well over fifty million of our citizens. This, by the way, amounts to substantially more than the *total* electrical consumption of our nation at the time of the Eisenhower Atoms for Peace address. It also approximates the amount of electricity normally consumed by those 50 million people who lost power during the August 2003 grid failure in the northeastern states.

Many other nations have followed the lead of the United States. Figure 41 shows the total amount of electricity generated by nuclear power in the top thirty nuclear-powered nations of the world in 2002. It also contains, as an insert, the percentage of electricity in these same nations derived from nuclear power in that same year. We note that the United States is still by far the largest user of nuclear power, but the United States ranks about number twenty among those nations most dependent upon this form of energy for producing electricity. Whereas the small nation of Lithuania ranks very high in its dependency upon nuclear power, it will soon lose this ranking because it will be shutting down one of its Soviet-built reac-

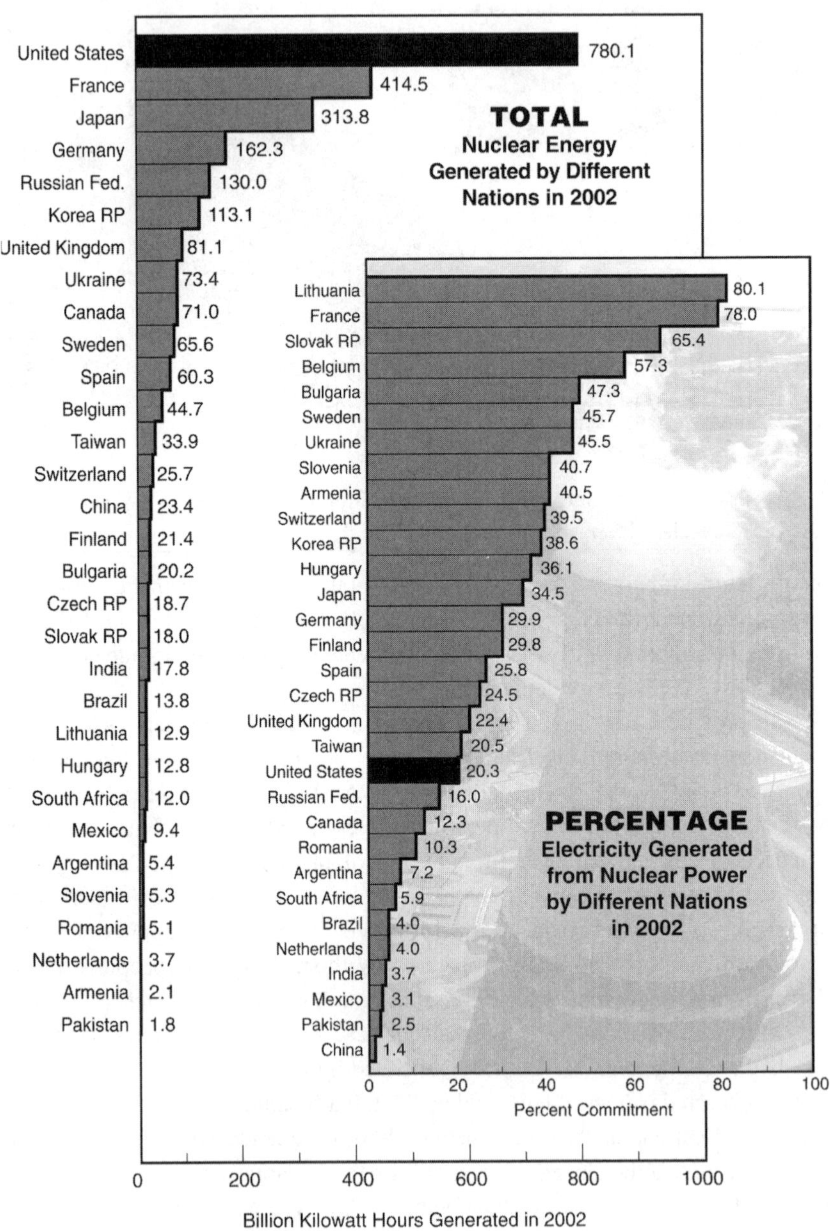

Figure 41. Total and percentage of nuclear energy generated by different nations in the year 2002. (Data courtesy of the Nuclear Energy Institute.)

tors in the relatively near future. France is among the most prominent nations in its commitment to nuclear power.

Ironically, Japan is more committed to nuclear power than the United States. Approximately one-third of all its electricity is generated by nuclear reactors. As we all know, Japan is the only nation in the world that has suffered the dark side of atomic energy (being the victim of the leveling of Hiroshima and Nagasaki by two atomic bombs to end World War II). Yet Japan, comprised of a small group of islands, is poor in energy resources such as oil, gas, and coal. This, combined with a large population and a strong industrialized economy, has convinced leaders in that nation to invest heavily in an energy source that is based more on technology than fuel—namely, nuclear power. The small amount of fuel required to produce nuclear energy is plentiful and easily imported.

The United States, on the other hand, has access to numerous forms of energy for producing electricity—such as coal, natural gas, oil, and hydropower—a luxury not common to many other nations. This, in part, explains why the United States is so far down on the list on its percentage use of nuclear power. As we shall see, more than two-thirds of the electricity currently generated in the United States comes from the burning of fossil fuels—which are neither long-term supplies nor compatible with global climate concerns. Hence, many energy planners are becoming even more interested in the future contributions that nuclear power might provide. It is true that the fuel for present-generation nuclear power plants, uranium, may be sufficient to power such reactors for only a few more decades at current rates of use. However, there are ways to substantially leverage the utilization of uranium, such that nuclear power could be viable for at least another millennium.

How It Works

Although an intimate understanding of nuclear power requires substantially more coverage than we shall even begin to attempt here, the basic idea of utilizing nuclear energy to produce electricity is simply to utilize the heat generated from fission and radioactive decay to build a furnace to boil water. As depicted in figure 42, the way electricity is produced in a coal-fired plant is by boiling water, using burning coal as a heat source, and then directing the hot steam to spin a turbine, which in turn spins a generator to produce the electricity. The high-pressure steam, having delivered its punch, leaves the turbine as low-pressure steam that is condensed back to liquid water (with the help of cooling water in a condenser loop). That water is then directed back through the boiler to complete the circuit.

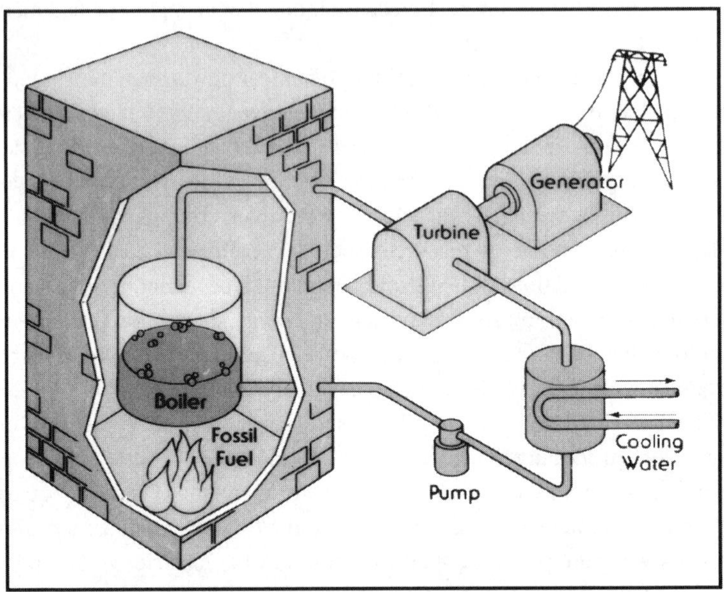

Figure 42. A coal-fired electrical generating station.

One common commercial nuclear reactor, called the boiling water reactor (BWR), is illustrated in figure 43. The only substantive difference between this power station and that shown in figure 42 is that the coal-fired boiler is replaced by a nuclear reactor.

The source of heat in a nuclear reactor is generated by the fission process. A fission takes place when a neutron is absorbed by a fissile isotope, such as uranium-235, causing the isotope literally to split apart, creating two or more smaller isotopes of other elements called fission products. Some of the very heavy isotopes, such as uranium-235 and plutonium-239, are called fissile isotopes because there is a high probability that a neutron can cause fission to occur if it strikes these isotopes.

The energy released in one fission event is about a million times that released when one combustible molecule is oxidized. The fission products created by this process move apart at very high velocities—their energy being determined directly by the conversion of mass to energy according to Einstein's famous equation $E = mc^2$. Here m represents the amount of mass lost during the fission event (the combined mass of the fission products is less than the mass of the original fissile isotope), c is the speed of light, and E is the energy imparted to the fission products. These fission products, though very energetic, are highly charged electrically because they lack several orbiting electrons. Consequently,

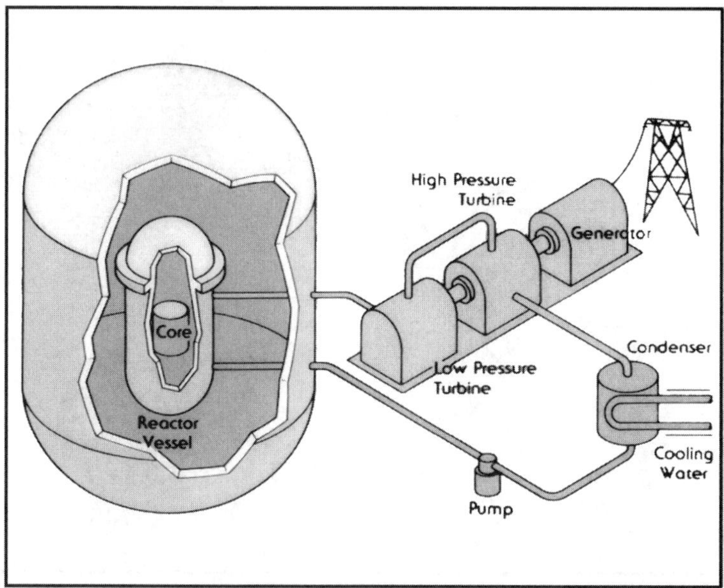

Figure 43. A boiling water reactor (BWR) nuclear electrical generating station.

they are stopped in the mass of material around them in a very short distance (a very small fraction of an inch), and their kinetic energy is transformed into heat—somewhat akin to the heating of brakes when a racecar is forced to come to a screeching stop. The heat deposited in the fuel matrix by the fission products and their subsequent decay is the heat that must be carried off by the coolant. The coolant (water) normally enters the bottom of the nuclear reactor and, in the case of a BWR, comes out as steam at the top of the reactor.

A most fascinating part of the fission process is that in addition to producing fission products, the fission event also yields two or three new neutrons. Some of these neutrons will be lost (they literally leak out of the reactor), and some will be absorbed in structural material and have no further value. However, in a properly designed nuclear reactor, enough of the neutrons will strike another fissile isotope (e.g., U-235 or Pu-239) to cause another fission event and keep this process going. This is called a chain reaction, as depicted schematically in figure 44. The trick is to achieve an exact neutron balance with time, that is, design the system so that for every fissile isotope destroyed by fission, there are just enough neutrons available to cause another fission to occur. If the neutron population decreases with every fission event, the power level will drop. Just the opposite is true if the rate of neutrons is increasing. Fortunately, this balance can be easily controlled, and once power reactors are raised to their desired power level, they can be operated at steady-state power for a very

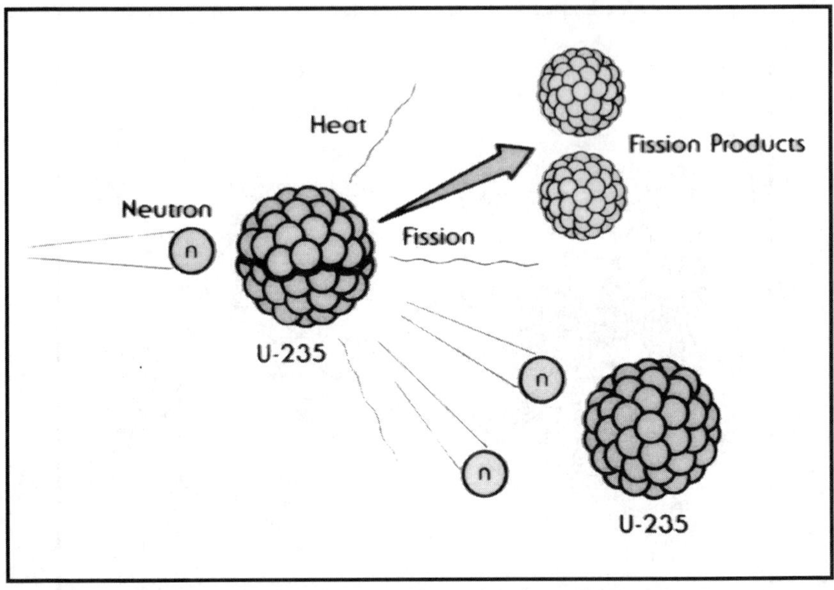

Figure 44. The nuclear fission and chain reaction process.

long time (months or years). At some point, enough of the fissile isotopes will have been destroyed (fissioned) that the reactor will need to be shut down for refueling.

Given this brief overview of the fission process, we note that the only major but important difference between a fossil fuel–burner plant and a nuclear power plant is the way that the heat is generated. Once the steam leaves the heat source, the remaining processes required to generate electricity are quite similar for both systems.

In both a fossil-fueled boiler and a nuclear power plant, the steam leaving the turbine must be condensed back to a liquid form (water), and this means some additional heat must be removed. This is accomplished by the condenser loop shown. Since heat must be ejected from this system, the utility operating the plant must decide what type of heat sink to use. If the plant is located near a river or lake, this heat can be diverted into a body of water. If not, it can be transferred into the air by a cooling tower—in a manner very similar to the air-blast condenser unit located outside of your home (which assumes that you have a heat pump or something similar as part of your heating and ventilation system to heat and cool your home). Since nuclear power plants tend to be very large, the amount of heat sent into rivers or lakes may be sufficient to alter water temperatures by more than acceptable amounts. Hence, air-cooling systems are often used. The large cooling towers often seen adjacent to a nuclear power

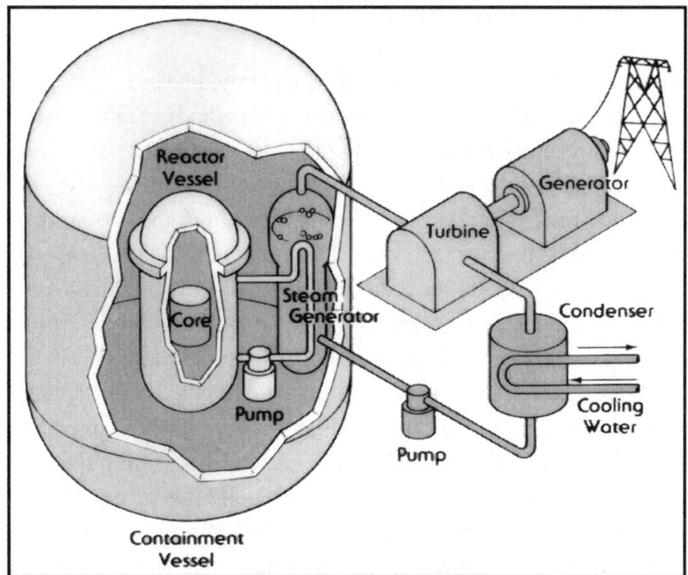

Figure 45. A pressurized water reactor (PWR) nuclear electrical generating station.

plant perform this function. Some nuclear power plants located near a large lake or an ocean can transfer heat into these huge bodies of water with minimal temperature changes.

Figure 45 illustrates the other commercial nuclear reactor in common use, namely, the pressurized water reactor (PWR). In this system, the water inside the reactor is maintained at a high enough pressure that it will not boil. Rather, it leaves the reactor at a high temperature still in liquid form and then is directed to a steam generator, where the lower pressure in a secondary loop allows the steam to be generated to drive the turbine. Beyond that point, the system is essentially identical to the others.

Whereas all the systems described above use water as the coolant, it is possible to use gas (such as helium), liquid metals (such as sodium), or even a molten salt to cool the reactor core. However, by far the bulk of all nuclear reactor systems in current use around the world for making electricity use ordinary water. One exception is in Canada, where they have developed the CANDU reactor that utilizes natural uranium and "heavy water" (deuterium oxide) as a coolant.

OTHER WAYS TO MAKE ELECTRICITY

Since the discovery in the 1800s that electricity could be useful for industrial and domestic purposes, there have been a variety of methods employed to make electricity. Nuclear reactors constitute only the most recent large-scale system.

One of the earliest methods was using falling water to rotate a water wheel connected to a generator. This system has progressed from the picturesque mountain stream waterwheel to the recent, massive Three Gorges Dam in China—where a 594-foot wall of Yangtze River water is directed through numerous penstocks to modern water turbines, resulting in the generation of some 18,200 MW of electricity (enough electricity to power either Shanghai or Beijing and a good deal more when fully completed in 2009). Hydropower, the name given to this method of generating electricity, is popular in regions of the world having large rivers that flow through substantial changes in elevation. Examples include the Pacific Northwest in the United States, Norway, and Austria, among others. Grand Coulee Dam, the largest in North America, is 550 feet high and generates about 6500 MW of electricity.

Advantages of hydropower include long-term viability, minimal atmospheric pollution, and low cost. Disadvantages include the diminishing lack of suitable new sites, loss of large land masses for farming or other productive uses because of the lake formed behind the dam, and the variability of the source—owing to strong seasonal variations in flow. Another concern is the affect high dams may have on salmon runs, since these fish must move through special devices around the dams during their travel from their freshwater birthing sites to the ocean and then back from the ocean for freshwater spawning. Although some increases in hydropower are feasible at existing sites, it is not likely that many new dams will be built, because of the lack of viable new sites and public opposition. Hence, the feasibility of obtaining substantial new blocks of electricity for hydropower is quite limited.

Fossil fuels provide the largest single source of energy for producing electricity. Coal, as mentioned above, is still the mainstay. The principal attribute of coal is its availability and relatively low cost. Huge deposits still remain in many regions of the world. However, the best low-sulfur coal is being rapidly depleted—requiring more-expensive procedures to mine the deeper and scarcer rich supplies. Also, transportation can become a major obstacle. China, for example, is thought to contain the largest deposits of coal of any nation, and these deposits are needed to serve the energy needs of its burgeoning population. Indeed, China is adding large numbers of coal plants for the production of electricity. However, the coal seams are mostly in the northwest region of the nation, far removed from the populous southeast region. The Chinese simply do not have the rail infrastructure capable of moving this source to the areas requiring power.

Though rare, it is possible for coal piles to freeze and become useless in the dead of winter—at precisely the time they are needed most. I vividly recall the winter of 1977 when I was teaching at the University of Virginia. The temperatures dropped well below freezing for nearly a month, even dropping below zero for several days. Several coal piles froze in the Ohio River valley—preventing the loading operations from feeding this material to the mouths of hungry furnaces. Nuclear power reactors in the region were widely credited with keeping the electrical grid energized.

But coal has several other disadvantages. One is environmental pollution, with both sulfur and nitrogen oxide discharges (leading to acid rain) and carbon dioxide (leading to global climate change). Another is public safety. Far more deaths occur each year from the use of coal (mining and railroad accidents, plus human lung diseases caused by atmospheric smog) than any other fuel source. Furthermore, it is a nonrenewable source—available for several more decades, but certainly not for the long term. Finally, coal is the chemical feedstock for numerous commercial products, ranging from laundry detergents to nylon hosiery. It is the key ingredient in the manufacture of iron and steel.

We need to question whether a depletable resource, such as coal, should be used in prodigious quantities to generate electricity when it is such a valuable feedstock for many other crucial elements of our economy.

Natural gas has become the recent "darling" for producing electricity in the United States. Ever since the California energy crisis of 2001, electricity producers have rushed to build natural gas–fired units. The advantages are that these plants can be built quite rapidly and inexpensively, and they can produce reasonably cheap electricity—when the price of natural gas is low. Unfortunately, natural gas supplies (along with the necessary pipeline network) are limited, and volatile gas prices have been climbing to the point that these new plants are being used only to provide "peaking" power—that is, only during those times of the day when there is the highest demand for electricity. Nuclear and coal plants tend to be "base-loaded," that is, they generally run twenty-four hours per day because they are less expensive to operate.

Natural gas is a cleaner fuel than coal because it produces no sulfur oxides and about half of the carbon dioxide of a comparable coal plant. However, it does emit nitrogen oxides. Perhaps of greater importance, gas is also a nonrenewable source of energy, a major reason for price variability. Furthermore, natural gas is a major feedstock to the chemical industry, which creates a mounting concern that an increased reliance on natural gas for electricity production could drive the price so high that many US industries would be forced to move to other countries. The ammonia industry is perhaps the most vulnerable, since enormous quantities of natural gas are required to produce fertilizers used throughout the world. Again, continued consumption of this ever-dwindling source for generating electricity needs to be examined through a lens focusing on economic factors and the depletion of resources.

Oil is another fossil fuel that has been used for producing electricity. It was one of the principal means of producing electricity in the United States in the middle of the last century. However, the OPEC oil embargo of 1973 sent a shock wave through the industry and caused a significant reconfiguration of the US electricity-generating system. Oil currently plays a relatively minor role in the production of electricity in the United States—producing only about 5 percent in 2002. However, oil is used in prodigious amounts for the transportation sector. Hence, we in the United States rely on oil for about one-third of our total energy consumption. Since we import more than half of this supply on an annual basis, this vulnerability is becoming more and more of a strategic issue among policy leaders, especially after 9/11. The real question is whether there is a way to offload our reliance on oil for the energy-intensive transportation sector. (We shall deal with this question in the chapter on transportation.)

I suspect many readers are wondering at this point, "what about solar energy?" "This is a free energy source, and it is inexhaustible!" Indeed, such pleas have been heard, and a great deal of research has gone into ways of harnessing energy from the sun, via methods such as direct solar furnaces, photovoltaic devices, and simply capturing the energy of the wind itself. Whereas some writers may draw a sharp line between "solar" and "wind" energy, I prefer to refer to both of these as forms of solar energy since they both rely directly on the sun as their driving force. Solar furnaces work on the principle of utilizing a large array of mirrors to focus direct solar rays onto a "point," where water is then boiled to make the steam necessary to drive electrical generators. Photovoltaic devices utilize an ingenious process to convert sunlight directly into electrical current. The wind, of course, is a direct result of the atmospheric pressure differences caused by the sun heating air masses at a different rate in various geographical locations.

A few writers may even consider hydroelectric power as a form of solar energy, since the rain cycle (essential to feed the flow of rivers) could not occur without the sun. However, I prefer to treat hydroelectric systems separately, mainly because they have far less dependency upon the timing of solar heating patterns. Neither solar furnaces nor photovoltaic cells can operate without constant access to the sun, and wind patterns need constant regeneration by daily solar heating. The time scale associated with hydropower is quite different.

While there are applications where wind and solar power can be ideal choices for heat and electricity, especially in remote and some rural areas, they remain far less desirable choices for the urban environments where large power sources are needed. The fundamental problem with all forms of solar energy is that it is a very diffuse energy source. We all intuitively know this because if it was concentrated like the sources we discussed above, we would not be able to walk outside. The maximum solar energy density on a midsummer day at high noon in Tucson, Arizona, is about seventy watts per square foot. That might not seem so diffuse if you happened to be the one sweltering in it, but it is the maximum that can ever exist. It diminishes significantly at other times of the day and year, becomes substantially reduced when cloud covers roam overhead, and disappears completely during the night. Hence, enormous collection areas are needed to capture the quantities of energy needed for modern civilization, especially given the low efficiency of converting sunlight to electricity. This means dedicating vast areas for mirrors, windmills, or other collection devices that cannot then be used easily for farming, dwelling, or other productive use. To put this into perspective, a typical 1,000 MW nuclear plant requires a land area from five hundred to a thousand acres. A solar complex with comparable electrical generation output requires about thirty-five thousand acres, and a similar-capacity wind farm requires about one hundred fifty thousand acres.[1]

The reason that wind farms require so much land mass is that wind has only about one-third the energy production capacity per acre of direct solar radiation, when practical considerations such as tower spacing are taken into account. Like direct solar radiation, wind energy is also quite variable, with only a few places on the globe offering suitable sites for reasonably reliable wind activity. Furthermore, the prime seasons for suitably harnessed wind energy are the spring and the fall, whereas the peak demand for electricity is in the summer and the winter.

Finally, it should be clear that we could never rely solely on solar power—even if we were willing to devote the large land collection areas necessary—without an immense energy storage system. Electricity demands continue around the clock, so solar systems could never do the job alone.

This might be a good place to comment on how national priorities and government regulations can substantially influence the prices, and therefore acceptability, of various energy sources. Wind has enjoyed considerable favor in the past decade as a renewable "solution" to our energy woes. Having grown up on a farm where the windmill was a source of pumping power for our water supply to the chickens, I can appreciate both the appeal and the nostalgia associated with this form of energy. Over the past couple of decades, wind power has benefited appreciably from substantial increases in size and efficiency because of huge government investments. But in addition to the research and development budgets allowed by our government to develop these devices (which now soar 250 feet above the horizon, with turbine blades some 300 feet in diameter), wind farms currently benefit from a 1.8-cent-per-kilowatt-hour tax credit subsidy. That alone reduces the cost of electricity to the consumer from wind turbines by about one-third.

To put that subsidy into context, in 2002 the total production cost of generating electricity from nuclear, coal, natural gas, and oil was 1.71, 1.85, 4.06, and 4.41 cents/kilowatt hour, respectively.[2] We note that in 2002 wind energy was *subsidized* by almost the exact amount of the total cost of producing electricity from nuclear power plants. One could justifiably argue that it is appropriate to subsidize new technologies to enable the technical progress necessary for competitiveness to mature. But given the barrier of the intrinsic low-energy-intensity property associated with all forms of solar energy, we should be advised not to allow ourselves to be duped into the fantasy that such sources will be truly economical—if judged on a level playing field—even into the distant future.

Biomass and geothermal are the remaining possibilities for generating electricity. Whereas many areas of the world still exist where wood and animal dung burning constitute a high percentage of the energy used for heating and cooking, biomass has rarely been proposed as a major future source for generating electricity if it were to rely on growing specific crops for creating the biomass. An enormous amount of land would need to be dedicated for such use. However, this low-technology option still has appeal for localized areas of the globe. Improved technologies may eventually allow significantly more biomass waste to be productively utilized. This would be a welcome development—consistent with environmental goals.

Geothermal energy is a term used to characterize the harnessing of heat residing deep within our Earth. Hot magma, resulting from the slow decay of radioactive materials that make up the bulk of our planet (yes, the bulk of the mass beneath the crust of our Earth is heated by radioactive decay!), sometimes protrudes close enough to the outer crust of Earth that it can be accessed. In fact, those who have visited Yellowstone National Park have likely witnessed gushing geysers like Old Faithful—made possible by the intersection of surface water with these naturally heated rocks. Hence, there are limited places where it is feasible to drill deep enough wells to pipe water down to the magma and tap the steam that is created to subsequently spin electrical generators or heat homes. The principal problem with this source is that there are very few areas where this is practical. Furthermore, naturally occurring minerals in the steam cause materials corrosion problems for plant equipment, along with environmental concerns. As such, geothermal energy will likely be limited for significant, long-term contributions to the generation of electricity. I had an occasion a few years ago to visit such a plant in New

Zealand, and the most memorable part of that visit was getting back into fresh, clean air. The sulfur "rotten egg" smell was ubiquitous.

We might mention that other schemes have been suggested for generating electricity. Tapping the energy of the oceans has natural appeal, since the tides continue to cycle on a consistent schedule. The idea here is to take advantage of the changing elevations of the water (somewhat similar to hydropower). Whereas this may one day prove feasible in some localized areas of the world (as it is today near Mont Saint Michel in Normandy, France), it is highly unlikely to result in more than a distant secondary source of commercial power.

Another highly touted source is nuclear fusion. Indeed, should it one day prove feasible to economically harness the energy of atoms being fused together, humanity would have succeeded in developing an energy source of an essentially unfathomable magnitude. The fundamental challenge of this process is that the fuel (deuterium or tritium) must be raised to extremely high temperatures, about a hundred million degrees, before the fusion process will occur. This is equivalent to the temperatures attained in the interior of the sun. Indeed, our sun is a huge nuclear fusion device, with the exceptionally hot nuclear plasma held together by the momentous gravity of the sun. Here on Earth, however, the gravitational field is much smaller than at the sun, and we need other devices to hold this plasma together. Since there is no material that has any structural integrity at such temperatures, the plasma must be kept away from the structural materials and held together by magnetic forces. Both the science and the engineering technology necessary to commercially harness this process are most challenging. Consequently, this energy source will almost certainly not be available for several decades, but the promise is so impressive that substantial research and development is continuing in most industrial nations.

Having summarized the energy sources currently used in the United States to generate electricity, figure 46 is included to help visualize these contributions.[3] We note that fossil fuel (coal, natural gas, and oil) constitute by far the largest source of our power, and they are all being depleted. Hence, there is substantial incentive to look for long-term supplies when we consider the future that we are leaving to our children and grandchildren.

Issues for Nuclear Power

Given the problems discussed above, why isn't nuclear energy in greater demand as new electrical capacity is needed? This is a complex question, but my experience suggests the answer is some combination of five factors: safety, waste disposal, nonproliferation, cost, and general public acceptance. I have attempted to

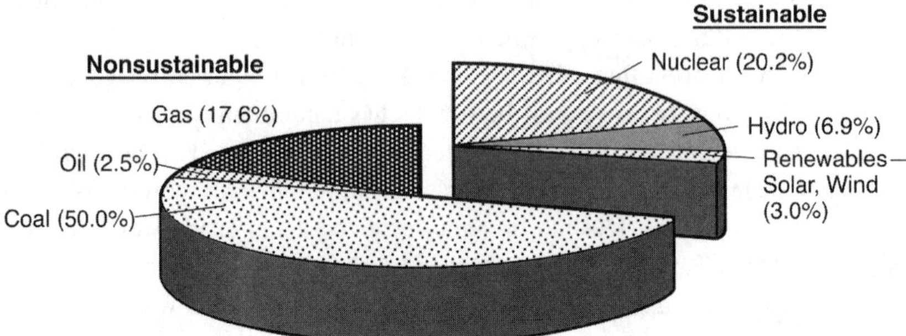

Figure 46. Energy sources for providing the electricity supply for the United States in 2002.

cover such issues elsewhere[4] in considerably more depth than can be allocated here, but I'll summarize some of the key points below.

Safety is certainly an important consideration, especially for a system capable of generating so much power in high concentration. This was, of course, recognized by the nuclear industry right from the beginning, and a great deal of attention has gone into both the design and the regulation of nuclear power reactors to minimize any harm to members of either the operating staff or the general public. Whereas it is not possible to guarantee absolute public safety (zero risks) from any meaningful activity, the record clearly shows that the production of electricity from commercial nuclear power plants is very safe. While the 1979 Three Mile Island nuclear power accident in the United States frightened many people, no member of the public has ever died as a result of a radiation-related accident at a nuclear power plant in the Western world. That's right. None.

There have been a few military personnel who died from excessive radiation while conducting sensitive experimental work at national laboratories during the early years following the discovery of nuclear fission. Likewise, there were thirty-one reactor operators and firefighters who died in the aftermath of the catastrophic 1986 Ukraine Chernobyl accident.* But the fact remains that no

* Of the thirty-one immediate deaths at Chernobyl, twenty-nine were firefighters. Although most of the immediate deaths were due to burns from the fire, some were directly the result of acute radiation exposure. Whereas there was wide speculation that hundreds or even thousands of people may die at some later time from radiation-induced cancer, we are now over fifteen years removed from this accident, and there are no confirmed cases of leukemia, which is the form of cancer most susceptible to radiation damage. There have been several thousand cases of childhood thyroid cancer, mainly in the Belarus region that was in the direct path of the airborne contaminants. Unfortunately, some of these children died, mainly because of poor postaccident intervention by health authorities. Fortunately, most were successfully treated. Tragic as this accident was, the number of actual deaths is far less than the speculations highly publicized by the international media. Still, substantial land masses were heavily contaminated, and it is only recently that some of these areas are being repopulated. This was, by far, the most serious accident in the history of nuclear power.

member of the public in any part of the Western world (including the United States, Canada, and western Europe) has been documented to have died of radiation-induced injuries resulting from commercial nuclear power. It is fairly well known by now that the Chernobyl reactor could never have been licensed in the Western world. In addition to some very basic physics differences, it did not have a containment surrounding the reactor system—a hallmark of all reactors licensed in the Western world. The roof of the building was of conventional construction.

There are numerous reasons for the impressive record of Western-designed nuclear reactors, including a defense-in-depth design philosophy that includes numerous physical barriers to be sure the high radiation levels inside the fuel system of an operating reactor cannot get outside the plant, even under severe accident conditions. Figure 47 illustrates the four primary barriers that prevent harmful doses of radiation from reaching plant workers or the public. First, most of the fission products created in the nuclear fission process stay lodged within the fuel matrix itself. The second barrier is the cladding, which is a metallic tube surrounding the fuel column. The third barrier is the primary coolant boundary. This boundary consists of a heavy steel pressure vessel (usually over six inches thick) and the associated heavy-wall steel pipes that carry the coolant to heat exchangers within a closed-loop system. Finally, all commercially licensed nuclear power plants in the Western world are equipped with an outer containment system. Whereas there are several designs that have been approved by the relevant regulatory bodies for such systems, a typical installation consists of a heavy steel-reinforced concrete structure (several feet thick), lined with an airtight steel plate of approximately one-half-inch thickness.

As a brief personal note on this point, I vividly recall a comment made by Erle Nye, chairman of TXU—a major utility in Texas that operates two nuclear reactors not far from Dallas. Erle was addressing a group of faculty and students at Texas A&M University at a special conference that I had arranged while serving as head of the nuclear engineering department at that institution. At that time, Erle was also serving as chairman of the Texas A&M Board of Regents. Since that event took place not long after the September 11, 2001, tragedy, his high-profile standing left him open to several questions relating to the possible vulnerability of nuclear power plants to suicide air crashes by terrorist groups. He first explained the features of reactor containment systems, including the concrete walls that are several feet thick and interlaced with massive amounts of reinforcing iron. Erle then calmly said that if he really believed a large commercial aircraft was hovering overhead piloted by terrorists looking for a vulnerable target, the place he would take his family for optimal safety would be to the

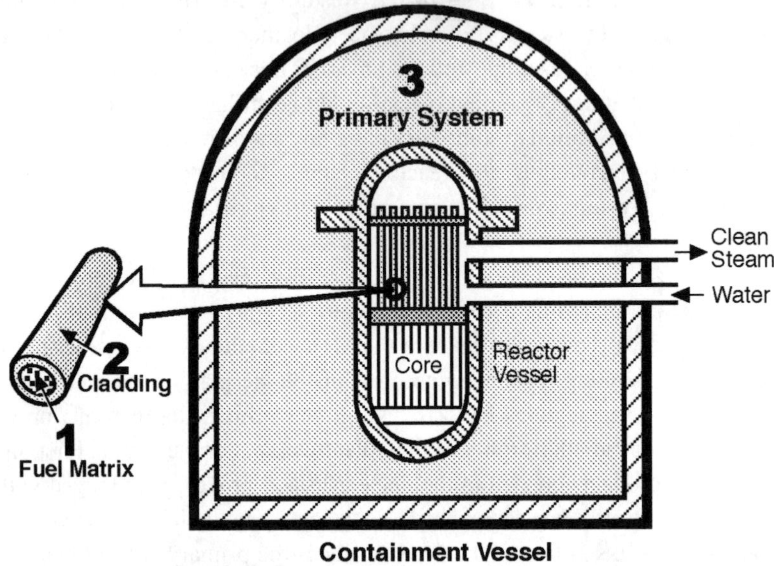

Figure 47. Physical barriers to prevent the release of radiation from a nuclear power plant.

inside of one of those plants.[5] There are few structures on Earth that have the strength of a Western nuclear reactor containment system.

Subsequent studies by safety experts have confirmed his confidence. Numerous calculations and scale-model tests clearly indicate that in addition to the exceedingly low probability of a giant aircraft being able to hit such a containment structure dead-center, even the engines would not be able to penetrate the exceptionally robust containment system should such a strike take place.[6] Whereas substantial damage could be done to equipment outside the plant (as would be the case for any large industrial complex), no parts of the engine, the fuselage, or the wings—nor the jet fuel—would penetrate through containment. One could always speculate on the consequences of a bomb being dropped on a commercial nuclear power plant. If the device were a conventional chemical explosive, it is hard to imagine the damage being any more severe than the direct hit of a massive airliner—the consequences of which were noted above. If it were a nuclear bomb (an incredibly low probability), the principal damage and associated hazard would be the bomb itself.

Whereas the admirable performance of nuclear power systems over the past half century has quieted all but the most cynical opponents of nuclear power on major safety issues, the question of what to do with the nuclear waste remains a highly controversial matter. At the heart of the issue is the fact that an enormous amount of high-level radiation exists deep within the fuel of an operating nuclear power plant, and this radiation does not all decay away when the plant is shut down or when the used fuel (often referred to as "spent nuclear fuel," or SNF) is discharged from the reactor. After a few hours, the radiation levels drop to less than 1 percent of that during operation, but it is a long time before the radiation levels drop to harmless levels. After the very short-lived materials decay away, most of the activity comes from the fission products strontium-90 (a beta emitter) and cesium-137 (a gamma emitter)—which both have half-lives of about thirty years. Hence, it takes about three hundred years (ten half-lives) for the radiation emitted by these two substances to have decayed to radiation levels below that of the original uranium that was mined from the ground to build the initial fuel.

But there are additional elements created in the fission process called actinides—very heavy elements (above uranium and plutonium in the periodic table of elements) that generally emit alpha particles. These alpha particles can be easily shielded (as noted in chapter 2), but they generate heat. More important, the half-lives of many of these actinides are tens of thousands of years. Hence, this material remains radioactive for several millennia, which brings up the issue of how to safely store this material for such long periods of time.*

It may seem that requiring storage for such an extended period of time is quite daunting, but we must first be reminded that the amount of this material is exceedingly small. We in the United States currently receive about 20 percent of our electricity from nuclear power. If we were to receive all of our electricity from this source for the next fifty years, the amount of space needed for the sequestration of the nuclear waste, resulting from this massive electrical generating capacity, would fit on a football field some fifty feet deep. Because of the heat still being generated in this fuel, however, it would be necessary to spread it out somewhat to allow proper cooling. Nevertheless, this is a miniscule volume—certainly a tractable problem, figuratively comprising only a tiny dot on our total landscape. This is also a very small fraction (well under 1 percent) of the waste generated from equivalent fossil fuel plants. Furthermore, whereas a long half-life of some several thousand years scares the uninformed, astute readers know that the longer the half-life, the lower the rate of radiation release—and the lower its intensity. Something with an infinite half-life is completely stable, that is, no radiation release occurs at all.

Because the amount of waste is so small, there are many ways that it can be safely stored. Indeed, all of the nuclear waste that has been generated to date at commercial reactors in the United States is stored right at the site—in specially

* We might note that many hazardous elements, including arsenic and several heavy metals, have *infinite* half-lives, that is, they never decay away. Yet they are in common commerce, remaining toxic forever.

designed storage units. In fact, one of the paradoxes concerning the nuclear waste issue is that there is no urgently pressing need to get it to a permanent geologic repository, which appears to be the most generally accepted choice for final storage. It can be stored on-site in a very safe fashion for a long time (alas, this cannot be said for the garbage that stacks up almost overnight in large cities during worker-strike conditions!). On the other hand, it becomes expensive to build new on-site storage facilities. Moreover, most professionals agree that it is best to have this material stored at a common site. Yucca Mountain, a ridge about ninety miles northwest of Las Vegas, Nevada, has been selected as this site in the United States. After a lengthy political battle, Congress and the president gave their approval to begin the licensing process at this site. Approximately twenty years of scientific work focused on Yucca Mountain provided the basis for this decision. If all goes well, this site should be open to receive nuclear waste packages by the year 2010.

I should hasten to point out, however, that not everyone agrees that the nuclear fuel elements discharged from our commercial nuclear reactors (SNF) is truly waste to be thrown away. Over 95 percent of the material discharged from the reactors is uranium, which can be chemically removed from the used fuel and recycled for use in the manufacture of new fuel. Furthermore, about 1 percent of the material is plutonium, generated in the fuel during the irradiation process from the transmutation of uranium. Plutonium is excellent fuel, and if the plutonium is mixed with a proper ratio of the recovered uranium, it can be put back into a reactor to generate much more electricity. The fission products can also be chemically removed to substantially reduce the heat load that is otherwise imposed upon the geologic repository. The point is that there is enormous value in the materials remaining in used fuel. Knowing how extensively radioactive materials have been used already for medical, agricultural, industrial, and other purposes over the past half century, I believe continued research and development will find ways to use beneficially most of what we now call waste. Hence, whereas I fully support the need for licensing Yucca Mountain—since there will always be a need to discard some nuclear materials—I hope that our nation will be wise enough to refrain from permanently burying SNF, thereby wasting an enormous potential for other beneficial uses. We have learned the environmental ethic of recycling in almost all other fields. It is time we in the United States employed that ethic to nuclear power, as is the case with several other nations, including France, Russia, England, and Japan.

Perhaps the most vexing issue associated with the question of recycling processed spent nuclear fuel for additional reactor burn cycles is that of nuclear proliferation. The concern here is that a nation or subnational group (including terrorists) might find a way to siphon off sufficient fissile fuel from a commercial nuclear power system to build a nuclear bomb. As we noted earlier, fissile fuel is material that will undergo nuclear fission when capturing a neutron (the principal examples are U-235 and Pu-239). Hence, fissile uranium and plutonium are the fuels of choice for making nuclear weapons. Plutonium is made in nuclear reactors, from a series of nuclear transmutations that take place in converting uranium into plutonium. To obtain top-grade weapons plutonium (so called weapons-grade plutonium, which is high in Pu-239 content), the fuel needs to be removed from the nuclear reactor after only a short period of time in order to avoid the buildup of unwanted plutonium isotopes, such as Pu-240 and Pu-242. These latter isotopes do not fission well, and they also produce spontaneous neutrons that would cause a weapon to fizzle, rather than to explode efficiently.

Commercial nuclear reactors, on the other hand, are designed to be operated for long running times—anywhere from one to two years before they are shut down for refueling. Furthermore, much of that fuel is "shuffled" to lower-power regions in the core of the reactor and then "burned" for substantially longer periods of time. Hence, "reactor-grade" plutonium is far less desirable in making a nuclear weapon (since such plutonium contains a high fraction of Pu-240).

Still, it is technically *possible* to build a working weapon with reactor-grade plutonium. Such a weapon would be far less powerful than one made from the proper ingredients, and there is a much higher likelihood that it would not work at all. But since it *could* work, at least well enough to have threat capability, there is concern that the commercial nuclear fuel cycle could be used to add to the number of nations (or group of terrorists) with nuclear weaponry.

It should be noted up front that no nation or group has to date used this route to acquire nuclear weapons capability. Seven nations currently acknowledge that they have nuclear weapons (the United States, Russia, the United Kingdom, France, China, India, and Pakistan), and a few others are suspected to have such capability (such as Israel and North Korea). These nations have all used either dedicated (non-electricity-producing) nuclear reactors to obtain weapons grade plutonium or they have built enrichment plants to acquire fissile U-235 (the other desirable material for making such devices). For low-technology groups, acquiring nuclear weapons via the U-235 route is considerably easier than pursuing the plutonium route.

If our concern is that new nations (i.e., those not currently having nuclear weapons) would use the commercial nuclear cycle to acquire such capability, we need to recognize that this is probably the most difficult path that they could choose. First, it is the most expensive; second, it provides the least favorable type of weapons fuel; and third, there is a good chance of getting caught—especially if they have signed the international Nuclear Nonproliferation Treaty (all but a handful of nations have signed it) because there are sophisticated international inspection systems set up by the International Atomic Energy Agency (IAEA) to quickly spot such infractions. Iran is an example of a nation under close scrutiny of the IAEA.

If our concern is that of subnational groups, such as terrorists, logic would suggest that they have far easier ways to invoke terror than trying to make a nuclear weapon. Despite what the popular literature might suggest, making an effective nuclear weapon is not something one does in the garage. There are several steps involved, requiring very specialized technical expertise and equipment—commodities that subnational groups don't have. Hence, it is substantially easier for such groups to steal a nuclear weapon already made than to go through the arduous task of building one, particularly since raw force is generally their forte. Furthermore, there are far-easier targets for instilling public panic, such as poisoning public water systems, setting off explosions during major sporting events, and so on.

If such groups wanted to use the word *radiation* to instill fear, they might be able to mix small amounts of more readily available radioactive material (obtained from general commerce) into a conventional chemical bomb and disperse it via a chemical explosion. Indeed, this has been widely publicized as a possibility; the so-called radiation dispersion device (RDD) or dirty bomb. We should quickly point out that such a possibility has very little to do with the commercial nuclear power sector, since radioactive materials are in common commerce in the medical and industrial sectors and far easier to obtain. Also, there is far less actual public danger from such devices than we might be led to believe. If the amount of radioactivity were large enough to constitute a real public hazard, the perpetrators would likely die first in the mere handling of the device. Hence, the value of an RDD to terrorists is more to create fear and panic rather than constituting a real health hazard.

Perhaps the biggest long-term hurdle for commercial nuclear power to overcome is its cost for construction of new plants. For the near term, the nuclear power reactors currently operating in the United States are very cost effective—delivering electricity to the grid at a cost just slightly lower than coal and much cheaper than oil and gas-fired plants. Only hydropower is generally less expensive. Figure 48 illustrates the trends of power cost via these conventional sources over the past several years.

Because of this low operating cost advantage, the utilities now operating nuclear power plants are doing very well financially. This performance has been achieved by exceptionally good management practices, where well-disciplined teams have dramatically increased the plant capacity factors (the fraction of time

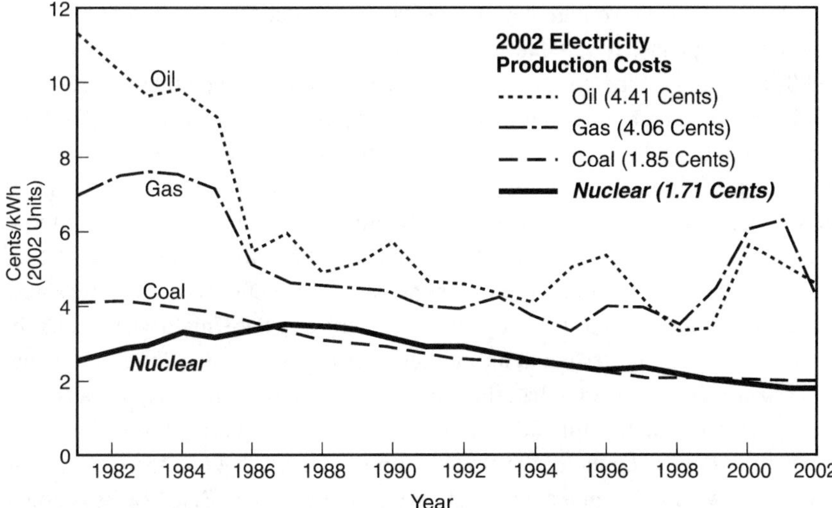

Figure 48. Electrical production costs in the United States from the current major energy sources.[7]

that the reactors are operating at full power) over the past decade. Also, the power ratings of many plants have been increased because of the enhanced understanding attained by such successful operation. This combination has increased the electrical output of the US nuclear reactor power fleet by the equivalent of about two dozen plants since 1990. Put differently, the United States has added the equivalent of bringing two dozen new plants on-line in the past decade, even though no new plants have physically been built. It is strictly a contribution achieved by the enhanced performance of the current fleet.

Given such improved and excellent performance, most utilities would like to keep their plants running longer than their licensed forty-year lifetime. Hence, as of the year 2004, several applications for an operating-license extension to sixty years have been made to the Nuclear Regulatory Commission (NRC), and approximately two dozen have already been granted. It is expected that a very high percentage of the 103 plants now on-line will be petitioning to extend their operating licenses for another twenty years of service.

Nonetheless, there will come a point when new plants will need to be constructed. The Nuclear Energy Institute predicts that the United States will need the equivalent of about fifty new plants on-line in the 2020 time frame just to retain the 20 percent market share currently held by nuclear energy. If environmental regulations are enacted to reduce the amount of electricity that can be provided by fossil-fueled plants, because of their undesirable emissions, the demands for new nuclear power plant construction will be considerably higher.

The challenge is to find ways to substantially reduce the construction cost of new nuclear plants. It was the high cost of construction during the 1980s and 1990s that brought the construction of new nuclear plants and other large base-load plants to a halt. Fortunately, new nuclear power plants have been designed that use considerably fewer materials than the present power reactors, and they are designed with even more robust safety than the current fleet. The result is that new construction costs appear reasonably attractive. Perhaps of equal or more importance, the licensing process has been streamlined to avoid the long delays that hampered construction progress in the past couple of decades. Whereas new plants were taking ten to fifteen years to construct in the United States, largely because of public protests and an inefficient licensing process, the same plants were being built in Japan in less than five years. Such time frames make an enormous difference in the ultimate cost of the plants and the willingness of investors to wait for a return on their investments. It now appears that commercial nuclear power is poised to begin a new era of service in the United States, as it is doing in many Asian and developing countries.

The final major barrier to the increased reliance on nuclear power is public resistance. In actual fact, a substantial majority of the American public supports nuclear power. However, in our free democracy, a small but committed opposition group can easily use the media to transmit fears, thereby exerting a negative influence far beyond their numbers. A high degree of self-fulfilling prophesy can be accomplished by a committed group opposed to new construction. By forcing litigation that doubles or triples the time of construction, the cost becomes so high that the plant indeed becomes an economic disaster—as predicted by the demonstrators! This is an unfortunate reality that few members of the public understand.

Yet, as implied above, national public opinion toward nuclear energy is quite favorable. National surveys by Bisconti Research, Inc., which have tracked public opinion toward nuclear energy several times a year for two decades, have found increasing support since the mid-1980s.[8] About two-thirds of the public favor the use of nuclear energy and about 80 percent support license renewal for nuclear power plants that continue to meet safety standards. The irony is that many people don't think their neighbors feel the same way. While a majority of a nationally representative sample says they personally favor the use of nuclear energy, the same polls often find that few believe a majority of others hold this view. A majority believes that others are either opposed or unsure. During the period around mid-2001, when there were numerous news stories about the important role of nuclear energy in the national energy plan and as a possible solution to problems such as the electricity shortages that were occurring in California and elsewhere, the public began to see nuclear energy's popularity in a different light.[9] Fewer thought that others in the public were opposed.

The Bisconti Research tracking surveys also show that support for building more nuclear power plants fluctuates depending on the perceived need to build more power plants in general. During the mid-2001 time frame, owing to electricity shortages, support for building more nuclear power plants peaked at two-thirds of the public but gradually returned to about half the public after electricity and energy problems became less salient. Support is generally higher close to nuclear power plants where plant performance has been excellent; the public is familiar with many people who work there, and the plants contribute significantly to the local quality of life.

It is also clear from these surveys that many people support nuclear energy simply because they believe it is the fuel of the future and not because of a real understanding of the benefits. In particular, many are unaware of the environmental benefits. Even a single sentence of informing the public that nuclear power plants do not emit any greenhouse gases or other air pollution increases the favorable percentage by at least 10 percentage points. The same polls find that the public cares very much about reliability, energy security, and clean air. With nuclear energy, all are possible. But even with such support, there is still a substantial reluctance to move forward aggressively. Why?

Given the hectic nature of twenty-first-century life in the fast lane, we have so many issues demanding our immediate attention that it is sometimes difficult to carve out the time necessary for long-term planning. Energy suffers badly from this point of view. As long as the lights come on when we throw the switch, we simply don't worry about where that electricity is coming from.

But it is imperative that we do so if we want it to keep coming. The rolling brownouts suffered by Californians at the turn of the new century provided an initial wake-up call. The recent grid failure in the Northeast that plunged fifty million Americans into the dark may have provided an encore (even though the principal problem in that event was grid stability, rather than energy supply sources).

From a global perspective, the September 2003 blackout in the entire nation of Italy is likely to bring a sobering reality. Italian officials may rethink the wisdom of having chosen to import the bulk of their electricity. Italy reacted to the Chernobyl accident by banning nuclear power production in that nation. Hence, they now import most of their electricity from Switzerland and France. A quick glance back at figure 41 reveals that Switzerland generates about 40 percent of its electricity from nuclear power plants, and the comparable number in France is nearly 80 percent. Consequently, the reality is that Italy is currently very dependant upon nuclear energy—even though it is a "nuclear-free" country!

Whereas the above discussion may help provide us with a bit stronger basis for reflecting on the alternative ways we have to generate electrical power, we still might ask if we are really going to need much more in the future than we have now. Why the worry?

It is within that context that I suggest we take a quick peek at history and then reflect for a few moments on the realities facing us in the future. Figure 49 contains a world plot of energy use from 1850 (when Abe Lincoln was still splitting logs to keep warm) up to the present, along with projections through the end of the present century. The data over the past 150 years is actual.[10] We see from this figure both a recent, rapid increase in world population and a rather substantial increase in the per capita energy use by citizens in the developed world. The question is, how will this pattern change in this century?

The three projections included from 2000 to 2100 in figure 49 were determined by a special committee of the International Nuclear Societies Council,[11] an organization made up of all the major professional nuclear engineering societies in the world (including those from Europe, North and South America, Asia, and Australia). The committee based its studies on data provided by the World Energy Council, the World Health Organization, and other highly reputable bodies. With the uncertainties associated with any predictive analysis, the three curves were offered in the spirit of bracketing expectations.

If we take the middle curve, which is deemed to correspond to the scenario of highest probability, we note that by the year 2050, the world population will have increased from six billion in the year 2000 to approximately nine billion. The total energy will have increased from 400 EJ to 1000 EJ.* Table 1 is included to illustrate these figures clearly. Whereas some will argue the numbers are too high and others will argue they are too low, it is sobering to note that today, US citizens are using nearly three times the energy per capita that table 1 indicates the average world citizen will be consuming in 2050. This is a staggering observation! When I first became aware of this, it hit me squarely in the solar plexus. To suggest that we don't have to take these numbers seriously is saying that the two billion members of our planet who have never seen a lightbulb need to be frozen in that position for the next five decades. Given the increasing speeds of transportation and the vast communication network opened by the Internet, it is clear to me that billions of people around the globe are becoming aware of the conveniences of life that abundant energy is providing to the Western world, and they want to be a part of it. To deny them this privilege, either consciously or unconsciously (i.e., by ignoring the issue), is simply unconscionable and, moreover, could lead to wars over energy that we have only begun to experience.

---

* 1 exajoule (EJ) is 10^{18} joules, approximately equal to one Quad, which is 10^{15} British thermal units (BTUs).

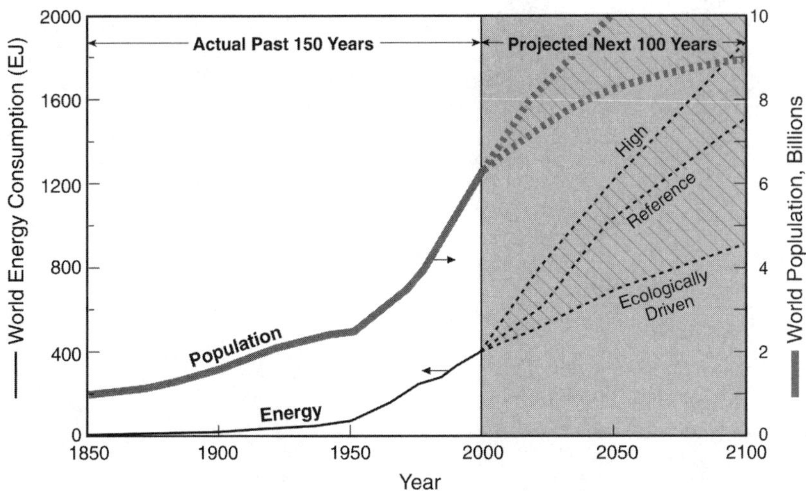

Figure 49. The huge rise in world population and energy use.

The consequences are enormous. Some will argue that through increased conservation, we can ward off the need for more energy—and more electricity, in particular. Certainly there are still many areas where increased efficiency is needed, and we should strive hard to achieve such efficiencies. But conservation will not *create* cool refrigeration, nor will it *create* light or heat. The world will need more energy in the future—considerably more energy—and this must be generated and delivered in a manner consistent with conservation and environmental stewardship. In many ways, it appears that nuclear energy came in providential timing.[12] Another great contribution made possible by the pioneering work of Marie Curie.

Table 1. Stark realities of the growing energy dilemma

Year	Population (Billions)	GJ/Person	Total Energy EJ
2000	6	67	400
2050	9	110*	1000

* US Today ~ 300 GJ/person

100 GJ represents five times increase for poor nations

7. Modern Industry

I have been wearing contact lenses for the bulk of my adult life. As a confirmed jogger, I find them much superior to glasses—especially during winter. I just don't like sloshing through rain and snow with glasses that have such a propensity for fogging up from the condensed vapor of my constant huffing and wheezing. Contacts also provide peripheral vision, something lacking with eyeglasses. But like many contact disciples, I sometimes get to the point where my soft lenses get so dirty that I need to submit them to a "professional rub."

It was during one of those occasions that I got more than an eyeful of refreshing clear sight. After asking Rob Young, my optician, if he could once again remove the grit and stains from my contact lenses, I asked if radiation was used in the optics business. He laughed and replied in the negative. He explained that he was not antinuclear, but opticians relied on the world of physics—not radiation. Satisfied with his answer, I then settled back in the chair as he "ground" away on my lenses with his special cleaning agents, and we began to solve the world's problems on a variety of topics.

Finally, as my newly refreshed contact lenses were bathing for their final soak, we drifted into a conversation on economics. Rob sheepishly revealed that he once made a mistake on an order for saline solution, which he routinely kept on hand for the convenience of his customers. He inadvertently misplaced the decimal point on the order, and he was delivered ten times the amount he had intended. Because of the size of the order, it came directly from the factory, rather than the supplier warehouse. He noted that the cardboard outer box contained a reference to radiation on its top and sides.

The word radiation? My ears immediately perked up. "Why on earth," I asked with bated breath, "was the box branded with a reference to radiation?" "Well," Rob replied as if this were a silly question. "I thought you knew that saline solution is irradiated with gamma rays to be sure there are no parasites. After all, you put this solution directly in your eyes. You want to be sure it is completely sterile, don't you?"

I almost fell off the chair. It all made immediate sense to me, but this was in direct contrast to the initial answer that Rob gave me when I asked if radiation was used in the optics business. The fact that radiation was used to sanitize the saline solution I used every day was so obvious to Rob that he just took it for granted. But it was news to me, and it once again confirmed how much the field of radiation has entered our everyday lives. As we continued our discussion, we readily noted that his patients who wear glasses also benefit directly from radiation, because a crucial step in making glass is to assure the proper amount of moisture is present during the manufacturing process of good glass. Neutron probes are routinely used to be sure that the moisture content in the molten sand is optimal for such purposes. Even eyeglasses made with plastic lenses may benefit from radiation, as we shall see.

That visit provided further impetus for me to continue my search to see how pervasive radiation is in modern industry. The comments below certainly do not cover all aspects of applying radiation techniques for product development. Rather, they just scratch the surface. But perhaps this introduction will help us to have a better appreciation for just how many of the products that we take for granted would not be available at the quality and low cost we have become accustomed to without the technology resulting from Marie Curie's marvelous discoveries.

Indeed, it is difficult to comprehend fully the enormous impact the harnessed atom is having in modern industry. Recent studies revealed that in 1995 the application of radioisotopes to US commerce resulted in over $300 billion to the United States gross domestic product, and this supported nearly four million jobs.[1] This was exclusive of commercial nuclear power. Whereas this did include agriculture and medical applications, a very significant amount of it derived from industry.

Though the modern factory furnishes us with most of the products related to using radiation that we make use of on a daily basis, most citizens are completely oblivious to this connection. However, once informed, when we think about the number of products now available, and the quality of products achievable, the correlation between the growth of the radiation industry and innovative product production becomes more believable.

PROCESS CONTROL

Modern manufacturing succeeds when products can be turned out in high volume, in high quality, with low cost. This places a high demand on instrumentation that can measure and rapidly adjust for any variations from product specifications during production. Quality control is essential for modern manufacturing when parts come from different companies and often different parts of the world. Typical measurements that are often made in production lines include liquid levels, the density of materials in vessels and pipelines, the thickness of sheets and coatings, and the amounts and properties of materials on conveyor belts.

Because radiation has the ability to penetrate matter, industrial measurements can be made using radioisotopes without the need for direct physical contact with either the source or the sensor. This allows on-line measurements to be made, nondestructively, while the material being measured is in motion. For a thorough and excellent technical coverage of the ways that radiation technology is employed in the industrial world, I highly recommend that the reader spend some time with G. C. Lowenthal and P. L. Airey's *Practical Applications of Radioactivity and Nuclear Radiations*.[2] I am very grateful to my Australian colleagues for writing that book and for allowing me to use much of their material for this chapter. Paul Frame and William Kolb's *Living With Radiation: The First Hundred Years*[3] also includes a wide array of radiation applications, but mainly from a historical perspective.

Level gauges generally operate on the principle of attenuation. A radiation source is placed on one side of a container being filled, and a detector is located on the opposite wall. As illustrated schematically in figure 50, when the liquid rises to intersect the line between these two instruments, the filling material substantially blocks the beam source; then the signal seen by the radiation detector drops dramatically. An electronic signal is instantly sent to the control station signifying that the container is full (or has reached the desired fill level). This simple but very fast and accurate device is often used in bottling plants to guarantee the proper filling of common soda cans. It is essential when the fluids are extremely corrosive as in many chemical plants.

I used to ask my students how they would feel if they stuck their quarter into a soft drinks vending machine and got a can of soda pop that was only half full. Their first response was to update me on some basic economics; namely, the vending machine would not respond at all until they chucked in at least three or more quarters! But assuming an acceptable economic man-machine negotiation, they, indeed, would not be pleased to have less than a full can of refreshing liquid. Suppliers, of course, are well aware of this fact, and they strive hard to

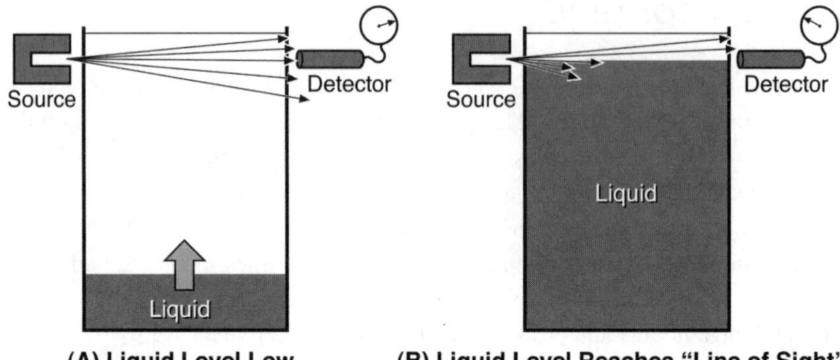

Figure 50. The use of radioisotopes to measure liquid levels.

utilize a foolproof methodology for assuring customers that their cans are filled to the brim before they are capped off and boxed for delivery. In most instances, this assurance is provided by the use of the radioactive sources mentioned above. The bottling (or canning) process must be accomplished so fast that total automation is a necessity. Radiation is ideally suited for such applications.

For more-complicated processes, like measuring the interface between two liquids that have similar densities such as different grades of oil, the backscatter (photon scattering) properties of gamma irradiation are often used. A gamma source is directed to the enclosure in question, and the intensity of the scattered gamma rays is measured at various angles to infer the location of the liquid-liquid interface. Microprocessors are sometimes used for these devices to process the data and enable a rapid determination. Such level gauges are widely used for monitoring or controlling the level of materials in tanks or hoppers in refineries and chemical processing plants to assure proper batch sizes for mixing operations. The data derived from such monitors are often fed directly to a central operating room to allow essentially instantaneous modifications of plant operations.

Radioisotope thickness gauges are unequaled in their performance and are used extensively in almost any industry involved in producing sheet material (e.g., sheet metal, paper, etc.). Sheet metal is used for making roofing, containers, car and airplane bodies, heating ducts, and so on. Paper, of course, is manufactured in a multitude of compositions and thicknesses for applications ranging from newspapers, magazines, and books to art prints and wallpaper. It is highly unlikely that rapid automation in such industries would be possible without the use of radioisotopes. Figure 17 (back in chapter 3) indicates how such a device works.

Modern steel mills utilize radiation gauges of this type to accurately measure the thickness of rolled metals at every moment of the production process. This is also the method used at paper mills, including the accurate measurement of wet pulp in the first stages of paper production. Most newer paper mills are equipped with radiation thickness gauges that will accurately monitor the thickness of paper up to speeds of over a thousand feet per second! This truly amazing capability has enabled quality paper to be manufactured very economically, thereby allowing a plethora of magazines and newspapers into our daily lives at very affordable prices.

Gamma rays are normally used for measuring the thickness of rolled metal, whereas beta particles can be employed for situations where penetration strength is not as important. Typical applications for beta particles include the paper, plastics, and rubber industries. The radioisotopes krypton-85 and promethium-147 are often used in such instances.

One of the items that we often pick up at the local hardware store is sandpaper. There always seems to something around the house or shop that needs a nice "rubdown" before applying that fresh coat of paint. Radiation thickness gauges are used three different times during the manufacture of sandpaper. The first application is in the making of the paper backing. This backing must be a precise thickness in order to have the strength needed, consistent with appropriate pliability. It is made in rolled sheets, as would be the paper used for the morning newspaper, but of course it is considerably thicker. The second step in the process for making sandpaper is to apply the glue. Here again, the thickness of the layer of glue is authenticated by radiation thickness gauges. Finally, the last step is to apply the grit. As we all know, there are several choices of grit size offered on the market, from very heavy (for removing the rough features of the job) to very fine (for finishing the project). Radiation thickness gauges are employed to assure the consistency of grit size.

Since beta particles cannot penetrate very deeply into matter, it is possible to use the property of beta backscattering to measure the thickness of thin coatings that cover a thicker structural layer. Examples include monitoring the thickness of the galvanized layer on rolled iron, the thickness of electroplated gold, and the thickness of plastic coatings on metals. The radioisotopes thallium-204 and nickel-63 are often used for such purposes.

It is often necessary to monitor the density of materials as a function of time during a manufacturing process or industrial flow system to assure proper mixing and flow of the constituents. Radiation density gauges are often employed for such applications—again because they are noninvasive, fast, and can be employed in relatively inaccessible places where harsh environmental conditions exist (such as high temperatures or extensive dirt and grime).

We note that density gauges are fundamentally different from level gauges, since the latter are essential on/off devices. Density gauges, on the other hand,

Figure 51. Sandpaper—made via a three-step radiation monitoring process.

often are required to detect differences in the density of materials as low as 0.1 percent (i.e., one part in a thousand). By selecting the proper radioisotopes (ones with the proper photon energy for the task at hand), it is possible to meet such requirements. Typical applications include the monitoring of slurry densities in pipelines as well as the monitoring of scale (corrosion) buildup on the walls of pipes. It is important to detect scaling deposits since they can significantly limit the capacity of a pipeline to transport material.

Another application of radiation density gauges is to assure the proper density of concrete grout. Concrete is used extensively in the construction industry, and the quality (ultimate strength) of the concrete is determined largely by the density of the grout during the time of the pour. Whereas this can be important in many civil engineering construction projects, it is especially critical in such installations as supporting off-shore oil platforms. Here the grout is pumped under pressure into the enclosed forms and continuously fills the space between the radioactive source on one side and the detector on the other side. Any anomalies in the grout density are immediately determined so that necessary corrections can be made. Meeting tight specifications for a construction project of this nature can literally mean the difference between success and failure of the entire project. Structural imperfections in the concrete footings and supporting walls could cause the entire platform to collapse into the sea.

A totally different application, although still related to underwater endeavors, is to continually monitor the potential for flooding of buoyant structural members of subsea platforms. Huge, air-filled tanks constitute one type of such a buoyant structure. If flooding of the tanks should occur, the platform could tip over and collapse. Hence, an early warning for water ingress in such crucial devices is tantamount to survival. Again, radiation density monitors are ideally suited for this use.

Closer to home, we might be interested in knowing that radiation density gauges are widely used in the plastics industry to improve the uniformity of products. There is likely not a household in the modern world that does not have a wide sampling of plastic materials (telephones, computer casings, light switch covers, plastic lenses for eyeglasses, etc.). Density gauges were used in all of these cases to assure a consistently uniform product. These density gauges are also frequently used in the food industry to ensure the proper filling of cereal boxes and other commodities used in daily commerce.

Plant Diagnostics

By being able to measure liquid levels, bulk densities, material mixing, and final-product thicknesses—all utilizing radiation gauges separate from the actual material flows—it is possible to provide excellent quality control in the overall processing plant. Signals generated from these radiation monitors can be recorded at frequencies as rapidly as every one to ten milliseconds, thus allowing for essentially instantaneous plant diagnostics to be made. Once any anomalous or out-of-spec behavior is noted, subsequent adjustments can be made for any crucial part of the manufacturing process.

In addition to the radiation measuring devices outlined in the previous section, a plethora of radioactive tracer techniques have been employed to investigate reasons for reduced efficiency in modern plant operations. As we noted in the chapters on medicine and agriculture, the tracer property of radiation allows us to determine product flow as a function of time deep within a system such as a processing plant. This information is crucial to allow the operator of a plant to know where critical machinery malfunctions may be taking place. With this knowledge, corrections can be made to optimize the throughput, as well as the quality and quantity of the product. Tracers are now routinely used to measure flow rates, study mixing patterns, and locate leaks in heat exchangers and pipelines. Typical radioisotopes used for such purposes in the modern industrial world are summarized in appendix B.

As an interesting historical side note, it was George de Hevesey who was the

Figure 52. Breakfast cereals, with proper fills assured via radiation gauges.

first to use radiotracers for industrial application. You might remember him as the young Hungarian lad who we introduced in chapter 4 as having used a small amount of radiation to prove that his landlady was feeding him leftover food. Well, a few years after this incident, while conducting his graduate research, George was asked by his mentor, Prof. Ernest Rutherford, the early radiation pioneer in England, to find a way to separate what was then understood to be "radium-D" (now known to be the beta emitter lead-210) from a mass of lead that had been derived from uranium deposits. De Hevesey conceived the idea of adding "radium-D" to bulk lead and then monitoring the beta particles to trace slightly soluble lead salts. It worked amazingly well—thus issuing in a whole new world of employing the tracer property of radiation for a multitude of applications. We have already reviewed numerous applications of radioactive tracers in the worlds of agriculture and medicine. As we noted earlier, Dr. de Hevesey went on to win a Nobel Prize in Chemistry for his insightful work.

There were several subsequent events that really enabled Dr. de Hevesey's discovery to flourish. First there was the 1934 discovery of artificial radioactivity by Frederic Joliot and Irene Joliot-Curie—the parents of Hélène Langevin-Joliot, who wrote the introduction to this book. At about the same time, Ernest Lawrence and M. S. Livingston designed the first cyclotron, a circular accelerator capable of manufacturing several radioisotopes. A further leap was accomplished by Enrico Fermi when, in 1942, he led a team to build the world's first zero-power nuclear reactor (very crude by today's standards) in a squash court under the stadium seats at Stagg Field in Chicago. It is because of modern accelerators and nuclear reactors that we now have the capability of producing such a wide array of radioisotopes for commercial use.

For many chemical, refining, and minerals-processing plants, the efficiency of the chemical process depends upon the contact times between different chemical ingredients. By attaching radioisotope tracers to the ingredients of interest, it is a relatively straightforward process to determine the residence time of each ingredient in the mixing process. Hence, chemical industries have become prime users of radioactive tracers. Sample products include paints, lacquers, and multiple grades of petroleum ranging from tar to aviation fuel.

It is true that chemical tracers (such as dyes) can also be used as tracers, and some are available in copious quantities. There is also an increasing use of mathematical models to simulate process flows, with some having sufficient modeling accuracy that the need for tracers is reduced. Furthermore, there remain occasional concerns expressed in some workplaces about the use of radioactive materials in their business. Nevertheless, radioactive tracers are still used in many cases to validate mathematical models and to monitor processes that are so complex that detailed analytical descriptions are simply inadequate. Fortunately, there is now a wide variety of radioisotopes available in numerous labeled compounds that are commercially available—sufficient for almost any task at hand (see, by way of comparison, appendix B). Also, radiation detectors are now so sensitive that radioactive tracers can be measured with a high degree of sensitivity, even if the tracer solution is strongly diluted. Finally, as noted earlier, a great advantage of radiotracers is that they are usually monitored external to the plant so that there is no interference with the product chemistry or production throughput.

The petroleum industry routinely employs radioisotopes on an international scale to locate leaks in oil or gas lines. For example, India recently completed an eighty-seven-mile-long crude oil pipeline and considered both a conventional approach (hydrostatic pressurization and visual inspection) and a radioisotope tracer technique to test for leaks. Indian officials chose the latter, which allowed testing to be completed in six weeks (relative to an estimated six months using hydrostatic pressure) and saved $300,000 in the process. The radioisotope tracer process was not only faster and less expensive; it also provided enhanced safety for the workers.

MATERIALS DEVELOPMENT

When we think of the manufacturing industry, perhaps the first thing that comes to mind is making iron and steel. This, after all, formed the backbone of the Industrial Revolution and later allowed the skylines of twentieth-century cities to take on an entirely new shape. The heart of the process of making iron is, of course, the blast furnace. It is there that the iron ore is turned into iron. There are three basic components of the blast furnace that comprise a successful operation. First, there is a shaft through which the iron ore, coke, and flux (limestone or quartzite) are introduced into the furnace. Second, there are the water-cooled shafts through which high-pressure air is blown into the furnace. Finally, there is the hearth at the base of the furnace from which the molten iron and slag are tapped. The efficiency of the conversion process from iron ore to pure iron depends highly on the quality of the coke (which supplies the carbon monoxide essential for the primary chemical reactions) and the efficiency of the hearth drainage. A proven way to make sure the overall system is operating at peak performance is to inject one radioisotope tracer into the air shaft on one side of the furnace and inject another into an air shaft on the opposite side. Gold-198 and cobalt-60 are good candidates for this use. During the tapping process, samples of the molten iron can be taken periodically (perhaps every few minutes) and assayed simultaneously for the two isotopes using a gamma ray spectrometer. If the furnace is operating according to specs, the response curves (i.e., the radiation signals) of both radioisotopes will overlap—reflecting symmetric residence times for the "boiling broth." If not, there are likely cold spots within the hearth and adjustments need to be made. Lanthanum-140 is often used in a similar manner to trace the movement of the slag.

In another sector of our modern chemical industry, high-energy radiation is currently employed to both form and degrade polymers. By bombarding this material with high enough energy to cause ionization (i.e., knocking off the outer atomic electrons), free radicals are formed. These free radicals can cause profound effects in polymers. It is possible to either enhance or degrade the polymers being irradiated.

In the case of polymer enhancement, the radiation establishes chemical bonds between adjacent polymer chains in a process called cross-linking. The products resulting from this process include polyethylene, polystyrene, PVC (polyvinyl chloride), and polyesters. We recognize these names as forming the raw stock for irrigation pipes, insulation, and a wide variety of household commodities such as carpets, draperies, and other domestic decor. They also enable fashion designers to create fancy clothing by providing special fabrics. Other products include conjugated diene butyl rubber and heat-shrink plastics.

Figure 53. Wires coated with "heat-shrink" insulation.

The process of making heat-shrink plastics is particularly fascinating. First a polymer, such as polyethylene, is formed into the desired end state and then cross-linked with either electron beams or gamma irradiation. Cobalt-60 is often used in the case of gamma irradiation. This irradiated material is then heated above its melting point and blown or stretched into a larger version of the same shape. This larger version is subsequently "frozen" by a fast cooling process. Since this polymer has the unique characteristic of a "memory effect," when next heated it reverts back to its original size and shape.

Such "heat-shrink" products are now widely used in the packaging industry. Wire and cable insulated with radiation cross-linked polyvinylchloride exhibits excellent resistance to heat and chemical attack. Such products are now commonplace in the automobile and aerospace industries, telecommunications, and in home electrical appliances. A high percentage of the wires that you might encounter if you removed the back cover of your TV set or lifted the hood of your car are products of these radiation procedures. This product is also used to make heavy-walled tubing to splice connectors in power cables and in the telecommunications industry. It is likewise often used as a sleeve to snugly protect oil and gas pipelines against corrosion. We also often see heat-shrink film used for food wraps and other types of packaging and even sometimes for books.

The above process is being used at an increasing rate to cross-link foamed polyethylene for thermal insulation and wood/plastic composites cured by gamma irradiation. The latter products are gaining favor for flooring in department stores, airports, hotels, and churches because of their excellent abrasion resistance, the beauty of their natural grains, and their low maintenance costs.

New radiation-induced polymer-based biomaterials are being synthesized in Poland to be used as dressings for burns, ulcers, bedsores, and skin grafts.[4] As one special example, a solution of polymers plus a seaweed derivative are irradiated to form a "hydrogel." These hydrogel dressings are compatible with human tissue. They adhere well to wounds and normal skin without stitching. They promote healing and can be removed painlessly.

In some cases, the irradiation of polymers can cause desired degradation, rather than cross-linking. This is called bond scission. Scission polymers include Teflon, cellulose, and polypropylene. One particular use of very high-intensity radiation (the order of 100 megarads) is to irradiate and degrade Teflon to the point that it can be ground into very small particles—sufficiently small to be used in printer's ink and as coating additives.

I learned about one unique application of radiation while on the faculty at Texas A&M. The university research reactor is used to "punch microscopic holes" in a roll of thin plastic sheets to make filter material. Neutrons from the reactor are guided to nearby "fission plates," pieces of very flat glass covered on one side with a thin layer of U-235. The neutrons induce fissions, and the fission fragments bombard a special plastic film as it is rolled between the two plates. This fission product bombardment provides a precise degree of damage in the plastic such that subsequent chemical etching can then open the exceptionally small holes sizes desired. These films have an amazing range of uses. One use of this material is to treat kidney patients by incorporating this special material into dialysis units. This unique material is also used for purifying water for ultrapure operations required in the computer chip–manufacturing industry. It is also used for ultrafiltration operations to separate bacteria from viruses in microbiology laboratories. These films can also be used as reverse osmosis membranes to make pure drinking water like you buy in the grocery store, to purify waste water, and even to make fresh water from ocean water. Who would have ever dreamed that kidney patients, computer chip manufacturers, microbiology researchers, and folks needing pure drinking water would be the direct beneficiaries of a nuclear reactor?

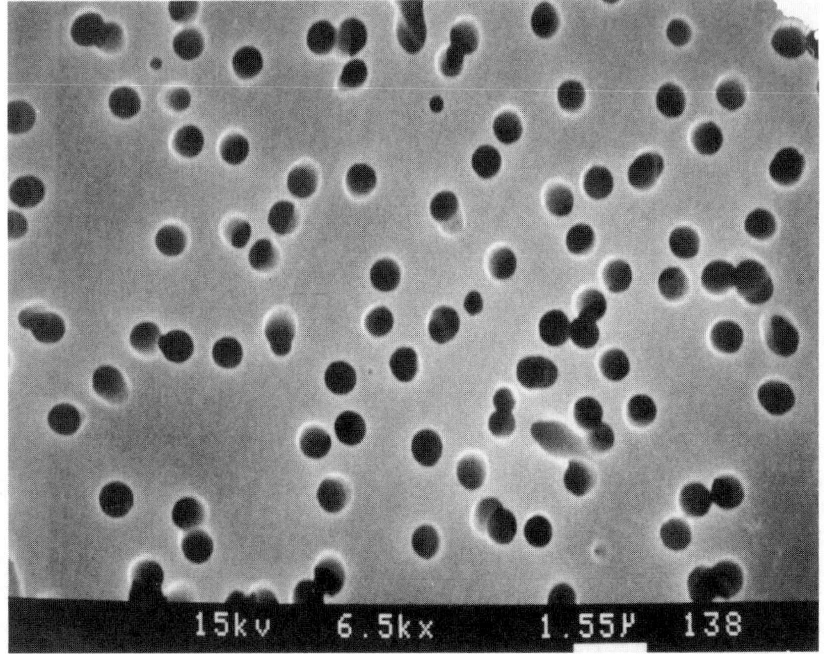

Figure 54. Plastic sheets perforated for kidney dialysis at a nuclear reactor. Note: these holes are about 0.7 microns in diameter, or about 1 percent of the diameter of human hair. (Compliments of Prof. W. D. Reece, Texas A&M Nuclear Science Center.)

MATERIALS TESTING AND INSPECTION

Anyone who has experienced the great pleasure of replacing an old toilet knows the "knuckle-bleeding" agony associated with trying to loosen the aging bolts and nuts that have been corroding via contact with impure water for unknown eons. Even the city water flowing through my forty-year-old household pipes is taking its toll, as mineral deposits continue to reduce the water flow at the far corners of my plumbing system. In the industrial world, such corrosion patterns can be ever more frustrating—given the chemical complexity of various fluids that are often used. One way to discern the degree of corrosion that has occurred inside pipes is to place a gamma source on one side of the pipe being investigated and a detector on the opposite side. By systematically moving this source-detector set of instruments up and around the pipe, clear images of the corrosion patterns can be instantly revealed, as illustrated in figure 55.

There is substantial interest within the world of materials research science in how cell structures contribute to the inherent properties of materials (such as

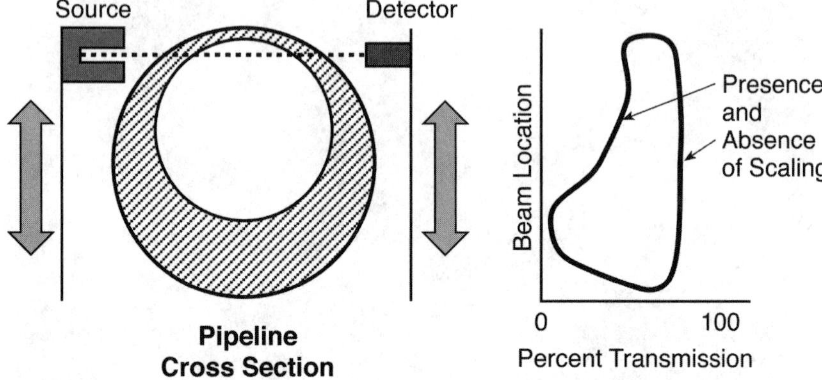

Figure 55. Determining corrosion patterns in pipes with radiation instrumentation.[5]

toughness, ductility, hardness, etc.). Standard optical microscopes can be of assistance, but they have very real limits on the degree of magnification that they can attain. Electron beams, however, can be used to substantially enhance the ability of materials scientists to probe the inner molecular structure of matter. This is called electron microscopy.

The fundamental difference between an optical microscope and an electron microscope is the mechanism of forming contrasts that result in producing the image. In an optical microscope, images result from differences in absorption of the light illuminating the object, whereas in electron microscopy they arise from the complex processes of electron scattering and diffraction. We noted in figure 12 (back in chapter 2) that high-energy photons result in higher frequency waves than those at lower energy levels. Electron wavelengths for the electron energies of interest are very short—shorter than average interatomic spacing. Hence, with modern high-resolution techniques, electron microscopes can resolve features at the atomic scale. Translated, this means that electron microscopes can achieve magnification levels up to about a hundred thousand. This allows researchers to obtain precision visualization of cell and grain boundary variations in a wide variety of complex materials.

The penetrating property of radiation is routinely used to check welds in crucial places such as airplane wings, housings for jet engines, and oil and gas pipelines. In many cases a gamma source is placed on one side of the material being inspected and a photographic film is placed on the other. Flaws can be readily detected on the exposed film, since a flaw will cause less attenuation than solid material, thus allowing more exposure on the film behind the flaw. With the advent of high-speed computational systems, new techniques based on photon scattering are now becoming available that are able to conduct these inspections without the need for film. For some types of inspections, it is preferable to use neutrons as the primary source of radiation, rather than photons. Californium-252 is a man-made heavy isotope that has a very large propensity for emitting neutrons. Hence, Cf-252 is often used to conduct inspections of critical welds on military aircraft. Another portable source for producing neutrons for these applications is americium-241 mixed with beryllium. This americium isotope emits an alpha particle, which then transmutes beryllium into carbon plus a neutron.

Other applications of the radiation penetration property include testing of nuclear reactor fuel assemblies (to assure the lack of metallurgical imperfections), detecting flaws in gas turbine blades, controlling the quality of ceramics, and confirming the presence of lubrication films inside gear boxes or bearings. This branch of inspection science, generally called radiography, has become quite sophisticated, allowing even three-dimensional imaging of objects under investigation. Capability of this type is called computerized tomography (CT). It is used in a variety of industrial processes, as well as in the field of medicine.

In a modern industrial CT system, more than a million measurements might be taken over 180 degrees to develop a crisp CT image. However, because of the advent of high-speed computational systems, this image can often be formed in a minute or less. For large-scale utilization, such as examining the cavity of a rocket motor, the x-ray source used may need to be generated by a linear accelerator capable of delivering energies up to fifteen MeV (fifteen million electron volts). At the other extreme, high-resolution microtomography systems can be utilized to complement the science of microscopy. Information at this microscale can be used to discern the cellular structure of materials such as polymers and wood, which would otherwise be destroyed during normal sample preparation.

Having grown up in a small farming community where many of our neighbors were employed in the timber industry, I was pleased to learn that methods similar to those described above are being developed to tomography uncut logs and mathematically determine the grain patterns that would appear on veneer sawed at any angle. This specialized system would allow sawyers to mill each log in a manner that would optimize the use of valuable timber.

Materials Composition

Many situations arise in modern industry where it is very helpful to know the composition of materials being considered for use as feedstock in a particular manufacturing process. In some instances, samples of the unknown material can be chemically analyzed to reveal the constituents. In other cases, however, it is most convenient to have a real-time, instantaneous analysis of the material flow. Here the property of x-ray fluorescence (figure 26 in chapter 3) can be employed. By focusing an appropriate beam of photons onto the moving assemblage of material to be investigated, inner electrons of the materials being investigated are knocked out, and the unique spectrum of x-rays emitting from the unknown materials can be measured to reveal the identity of their composition. An example of this technique is to assay the content of ore-bearing precious metals.

NONNUCLEAR ENERGY APPLICATIONS

We noted in the previous chapter that radiation is the key ingredient in the generation of electricity from nuclear power plants. However, only about 7 percent of the total energy consumed in the world today comes from nuclear power plants (approximately 17 percent of the global electricity generated, but only about 7 percent of total global energy consumed). Nonnuclear sources currently comprise the remaining 93 percent of the total energy utilized worldwide. By far the bulk of this nonnuclear energy comes from fossil fuels (coal, oil, and natural gas). It is very important, therefore, to note that radiation likewise plays a key role in improving the overall production efficiency associated with the burning of fossil fuels for electrical energy production.

The coal industry, which currently accounts for well over half of the world's supply of electricity, benefits directly from using radiation gauges to measure the moisture content in coal, determine the ash content, determine the content of other ingredients (such as sulfur), determine the BTU (heat content), and measure the rate of flow of coal into the furnace. In order to obtain optimal burn efficiency, coal-fired boilers must be fed with coal having a constant caloric value.

One way to obtain the above measurements is to bombard the coal entering the plant on a conveyer belt with cesium-137 and americium-241. The Cs-137 emits energetic gamma rays (662 keV), whereas the gamma rays emitted from the Am-241 are a factor of ten times weaker (about 60 keV). The high-energy Cs-137 gammas are attenuated by Compton scattering (a process essentially independent of the composition of the material). The lower-energy gamma rays from the Am-241 are attenuated by the photoelectric effect (which is very sensitive to the atomic number of the flowing materials), thereby easily revealing the presence of iron or silicon in addition to the desired carbon content. Hence, the flow rate of the coal being fed into the furnace is determined primarily by detection of the Cs-137 signal, whereas the percentage of ash in the mixture is determined by a computer analysis of the transmission of both the Cs-137 and the Am-241 gamma rays.[6]

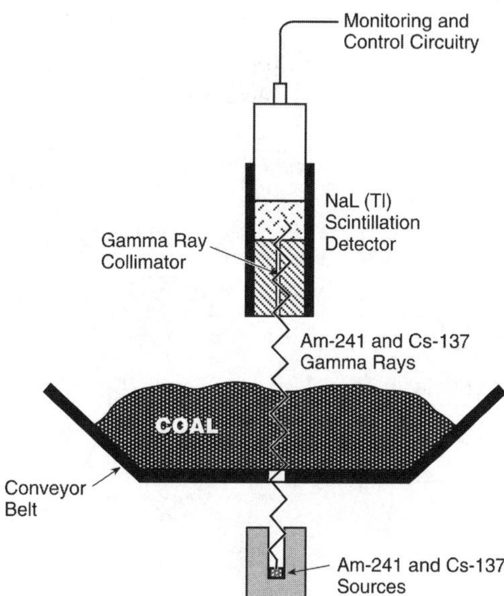

Figure 56. Analyzing the quality of coal while entering the plant on a conveyor belt.[7]

More-sophisticated radiation gauge systems can also be employed to measure other trace ingredients, including moisture content. As mentioned above, one approach is to employ the x-ray fluorescence technique mentioned in the previous section. Another is to employ a technique known as prompt gamma ray neutron activation analysis (PGNAA). Here the trick is to irradiate the coal along the conveyer belt with neutrons (from a portable neutron generator, such as californium-252 or the Am-241/beryllium system) and cause some of the materials in the coal stream to become activated (see, by way of comparison, figure 20 in chapter 3) thus releasing prompt gamma rays characteristic of the materials to be identified.

It is important to determine the sulfur and nitrogen combustion products because these gases are of considerable environmental concern in causing acid rain. Sulfur that resides in the coal entering the plant combines with oxygen and forms a variety of sulfur oxides. Nitrogen is always present because it is the major component of air. Hence, because of the very high temperatures attained in the coal furnace, nitrogen oxides are also formed.

It is within this context that one particular application of radiation technology is gaining increasing attention. Conventional environmental-protection devices, known as scrubbers, are widely used to remove large quantities of sulfur dioxide from the discharged flue gas. However, the by-products of this process have no commercial value, thus causing additional waste-disposal problems. Also, no reliable conventional process has been developed for simultaneous removal of both sulfur and nitrogen oxides in a single-stage operation.

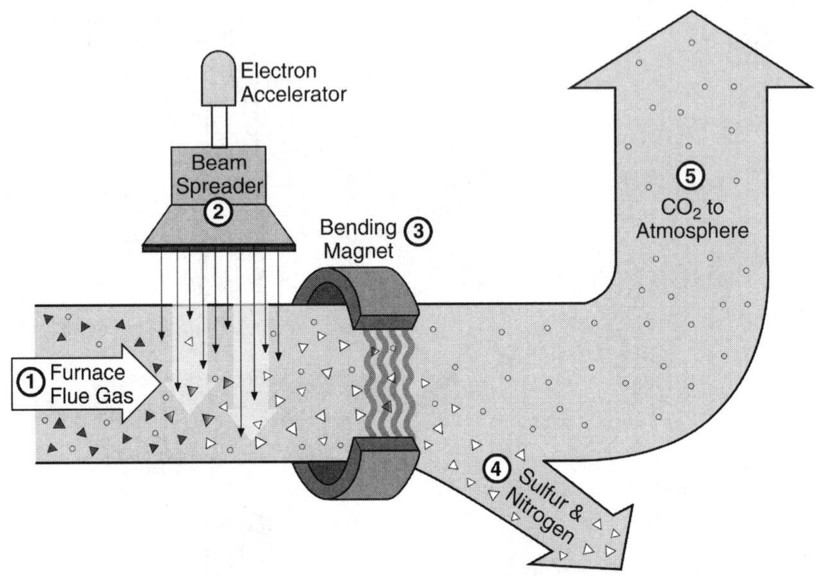

Figure 57. Facility to extract sulfur and nitrogen out of flue gases.

A new radiation technique, called electron beam (EB) processing, has been demonstrated to effectively remove both sulfur and nitrogen oxides in a single-stage process. I was privileged to tour such an experimental facility in Europe (the Pomorzany Electric Power Station in Poland)[8] in the mid-1990s. By passing the sulfur and nitrogen flue gases under a strong beam of electrons, these pollutants become ionized (hence, electrically charged) and can then be collected on positively charged plates or diverted by a magnet—substantially ridding the flue gas from these environmentally objectionable contaminants. An additional attribute of this process is that these toxic substances can be converted into a commercially viable agricultural fertilizer. This technologically demonstrated system has not yet been employed in many commercial coal-fired plants because of the economic considerations, but it may become more attractive if and when environmental restrictions become tighter.

The oil industry also is heavily dependent upon the use of radiation to conduct business. Finding new oil fields is a constant effort, especially as traditional sites are becoming exhausted. The standard way to determine the siting of new oil fields is to drill boreholes (wells) into the ground at strategic locations and then instrument these wells to make geologic measurements at various levels down the shaft. Borehole logging is a term used in conjunction with analyzing these test wells to determine the potential for economically viable oil deposits.

At least four methods based on radiation techniques have been developed to help determine which geologic strata being probed by the boreholes have the highest probability of being developed for oil production. One is based on gamma ray backscattering, in which energetic gamma rays are emitted by two gamma sources placed into the borehole. One emits high-energy gammas and the other emits relatively low-energy gammas. The detector system placed near these sources in the borehole can determine the bulk mass of the surrounding material from the intensity of the scattered high-energy gammas (due to the Compton scattering effect). It can also determine the constituency of the bulk (i.e., the specific elements in the matter surrounding the borehole) from the intensity of the scattered low-energy gammas (because of the photoelectric effect). The goal is to determine high hydrocarbon-bearing regions, characteristic of oil deposits.

A second method is to employ the prompt gamma ray neutron activation analysis (PGNNA) method outlined earlier. By identifying the carbon to oxygen ratio, a strong inference can be made regarding the presence of oil-bearing regions. The carbon to oxygen ratio varies from zero in the absence of hydrocarbons to a maximum value when the surrounding strata are saturated with hydrocarbons.

A third method is to utilize a neutron source in the borehole by employing the neutron backscattering principle to determine the hydrogen content. The americium-241/beryllium system yields neutrons in the 2 to 10 MeV range and can work well as an appropriate source. Neutron counters such as boron triflouride detections are sensitive only to thermal neutrons (neutrons that are thermalized most readily in a region of high hydrocarbons—because of the ability of the hydrogen atoms to rapidly slow down fast neutrons). Furthermore, they are essentially insensitive to the fast neutrons coming from the source (since boron has a very low neutron capture cross section for fast neutrons). Hence, such detectors can be placed very close to the source without difficulty.

Yet another use of fast neutrons (though perhaps more expensive) is to utilize a pulsed neutron source by employing a high-voltage ion accelerator. By accelerating deuterium onto a tritium target, 14 MeV neutrons are generated. Since the beam current (deuterium in this case) can be turned on and off in an accelerator, a fast beam of neutrons of only about ten microseconds in duration can be produced. A neutron detection system then measures the time it takes for backscattered neutrons to arrive, thus yielding a good estimate of the density of hydrocarbons in the vicinity of the region. I am pleased to note this application of pulsed neutrons, since I struggled for about four years at the University of California (Berkeley) doing my doctoral work using a deuterium/tritium source/target system (although for quite a different application). I'm sure the petroleum industry has been able to take advantage of a considerably more compact and efficient source than I had glued together for my dissertation.

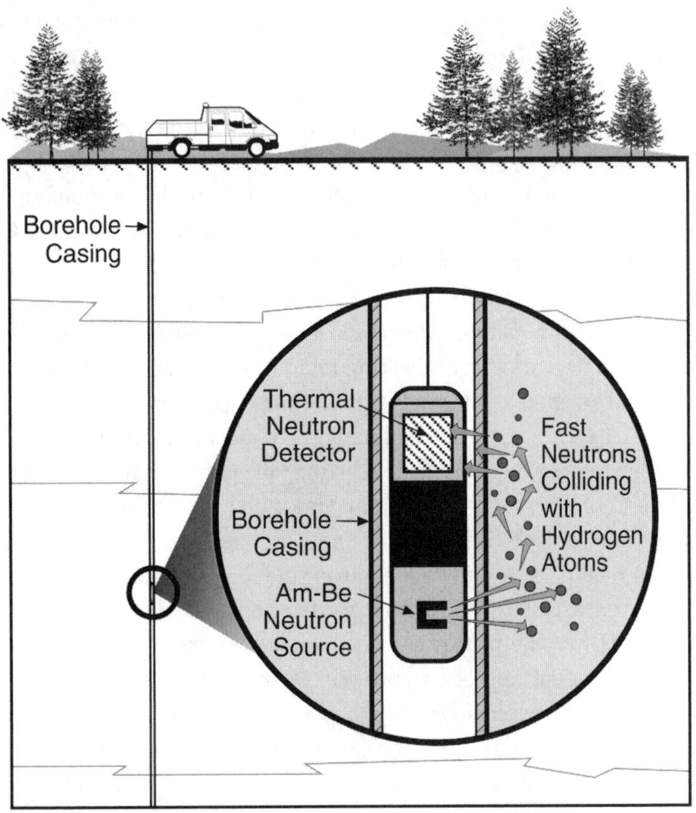

Figure 58. Borehole logging for determining prime oil well drilling sites.

Efficient refinery operation is also a very important part of the oil industry. Whereas it is difficult to install and maintain diagnostic probes inside the distillation towers (owing to the extremely harsh chemical and temperature environment), gamma probes can be rather easily installed on the exterior of the towers and then be moved up and down the tanks to record the composition of ingredients at various vertical levels. Any malfunctions within the tanks can be readily detected. Many of these distillation columns are very large (measuring some ten feet in diameter and ranging to heights of a hundred feet). Hence, being able to obtain accurate measurements of the contents is very important to the overall economics of the operation.

Neutron backscatter gauges are also routinely used in the oil refinery industry to monitor the levels of liquids in tanks and to determine the interface between different liquids. Because hydrogen is a highly efficient neutron moderator, the detection of backscattered thermalized neutrons can be used as an effective measure of the concentration of water or hydrocarbons in tanks.

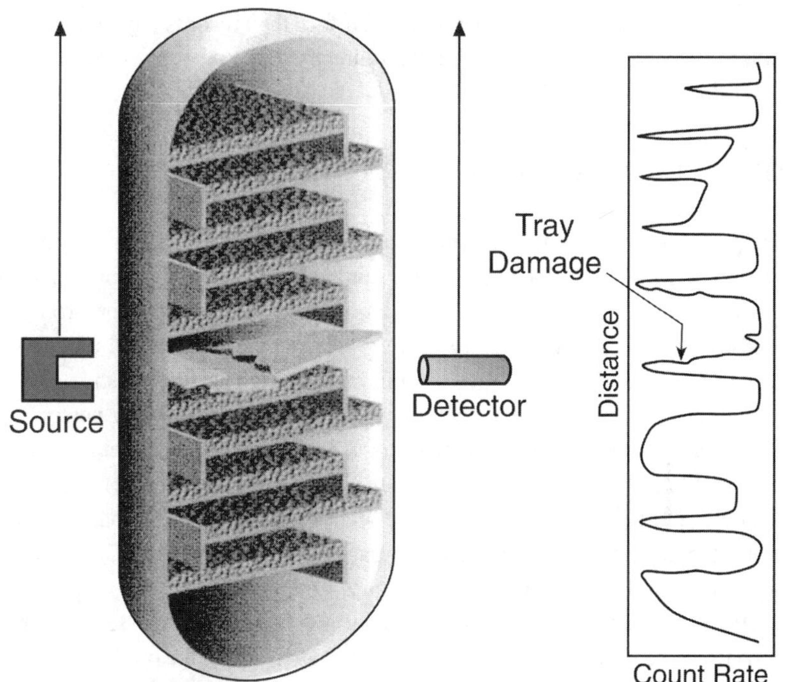

Figure 59. Use of radiation gauges to check oil refinery operations.[9]

CONSTRUCTION

The key to any successful construction project, whether it is highways, bridges, skyscrapers, or homes, is to assure a solid foundation. Substantial packing is absolutely necessary to ensure that the superstructure will not settle. One of the principal ways to provide solid footings is to use radiation gauges to determine both the density and the moisture content of the soil beneath the foundations.

Radiation gauges to perform this type of analysis normally use a gamma source such as cesium-137 or cobalt-60 to determine soil density. For soil compactness measurements, the gamma source is lowered into a small test hole and the density of the soil is determined by the number of gamma rays reaching the detector at the surface. If the objective is to measure the density of thin layers of asphalt or concrete in road construction, the gamma source is placed on the surface of the layered material and the density is determined by the number of backscattered gamma rays.

For determining moisture in the soil, a neutron source, either californium-252 or a mixture of americium-241 and beryllium, is normally used. Fast neutrons, directed into the soil from the surface, are slowed after colliding with the hydrogen atoms present in the soil. A detector at the surface then records the number of thermalized (slowed) neutrons, which relates directly to the amount of soil moisture.

PERSONAL CARE AND CONVENIENCES

We opened this chapter with an incident in which I learned about radiation being used to sterilize the saline solution for contact lenses. We also noted that those who wear glasses also benefit from radiation owing to the use of neutron probes in assuring the proper moisture content during the making of the glass, or through radiation density gauges used in making high-quality plastics.

But there are many other items we encounter that account for personal conveniences. Examples would include household bandages, wherein the gauzed part has likely been sterilized by gamma irradiation to assure proper sanitary standards and the thickness of the glue on the remainder of the band has likely been established by radiation thickness gauges. We noted earlier an extension of this process in the manufacture of ordinary sandpaper.

From an even-more-personal standpoint, women may be pleased to learn that most cosmetics have benefited from gamma radiation before the product is placed on shelves in the department stores. Whereas I have never been in a cosmetics factory, I am told that the ingredients and processes used somewhat parallel that of a sausage factory. The oils and greases required to achieve the desired colors and textures have a considerable propensity for attracting impurities that would not be appreciated on the human face. Fortunately, the final product is subjected to either beta or gamma irradiation to be sure that no live parasites remain by the time the product can be confidently used. Unirradiated lipstick just might contain a host of microscopic "visitors" that would not be appreciated under most romantic conditions!

One of the most useful features of radiation is to change the molecular structure of some materials to allow them to absorb huge amounts of liquid. Several industrial applications benefit directly from such transformed materials, such as air refresheners. Of more personal interest, this process is being used to manufacture very absorbent disposable baby diapers and tampons. Both products have enjoyed high consumer acceptance.

Certainly the wide variety of clothing choices that we encounter at the local mall adds immensely to our daily lives. One of the main benefactors of radiation, as we learned from chapter 4, is the cotton industry. Hence, we might be reminded that many of those new cotton styles would either cost a great deal more than they do, or possibly not be available at all, were it not for the huge progress made in cotton yields by radiation mutation technology.

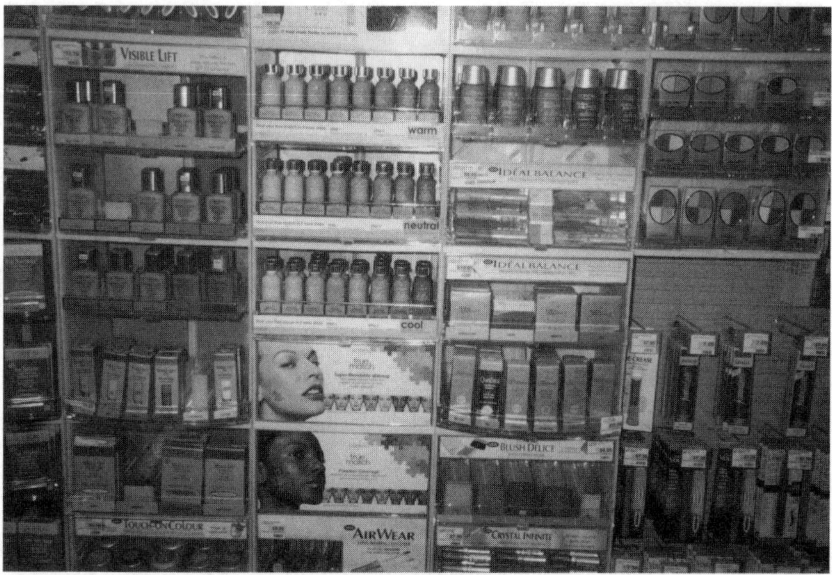

Figure 60. An array of cosmetics all sterilized by radiation to assure safe use on human skin.

Just one last example. The bottled-beverage industry, as we all know, is huge. Whether our quest is for that cold bottle of beer, a can of soda pop, or just plain bottled water at the end of a hot day, we can be thankful for radiation technology that allows these cans and bottles to be made at a very reasonable cost. Both glass and metal containers almost assuredly included radiation devices somewhere during the manufacturing process. And the automation invoked in the bottling process could never have been accomplished at such speed and low cost without the benefit afforded by radiation measuring devices.

Perspective

I mentioned earlier that one of the highlights of my tenure at Texas A&M University was to help organize a special series of programs celebrating the legacy of Marie Curie. Approximately a week before the Marie Curie Exhibition was scheduled to open at the J. Wayne Stark Gallery, curator Catherine Hastedt called me to say she was trying to develop a special "hands-on" display that would interactively pose questions to viewers. She wanted to display about a half dozen types of familiar products that we encounter every day and ask which ones do *not* benefit from radiation at some point during the manufacturing process.

Figure 61. Bottled beverages—all assured of being filled to the brim via radiation gauges.

I had never heard the question asked in this manner. Prior to this, I had been searching for processes that *did* include radiation in the production process, rather than ones that did not. As I pondered this question, Ms. Hastedt kept throwing examples my way, such as purchasing a can of soda pop, buying a piece of wooden furniture, and so on. We all now know that radiation is used at least twice for the case of soda pop. Radiation thickness gauges are utilized in the rolling of the sheet metal used to make the soda pop can and again during the filling process to assure the liquid comes right up to the brim before being capped. Many large furniture companies require proper drying of their lumber before they will accept it for making fine furniture—and the rate of moisture retained within the lumber is, therefore, often monitored by the use of neutron probes to assure appropriate rates of drying.

Cathy and I spent nearly a half hour before we could come up with a *single* industrial product that did not take advantage of radiation processing somewhere in its evolution from raw resources to the finished product. We finally discovered some type of a photographic process that, to the best of our collective knowledge, was based purely on chemistry. Perhaps an astute reader can even correct me on that account, but I think you get the point. The modern industrial world benefits enormously from the harnessing of radiation. And we, as consumers, are the fortunate recipients.

8. Transportation

The invention of the internal combustion engine constituted one of the most revolutionary events of modern history. A gaze through the window of any city building in America (and indeed, most of the Western world) readily reveals the grip that cars and trucks have on our everyday lifestyle. By the turn of the twenty-first century, there were approximately two hundred million cars and trucks on the road in the United States alone, and the numbers grow every year. Today it is hard to imagine how we could live without them. But how many of us realize the role that radiation plays in bringing about these modern modes of transportation? Let's go on a quick "drive."

Cars and Trucks

As a boy, I can vividly remember the belching smoke wheezing out of the tailpipe of road vehicles in the forties and fifties. Ring wear in our farm trucks sometimes got so bad that we had to hold our breath any time we needed to move behind the open bed of the vehicles to load hay bales from the opposite side. In addition to the obvious pollution created by the engine wear, considerable power was lost as combusted exhaust slipped past the pistons—rather than give them an extra boost.

I got another dose of bad engines while attending graduate school. During my senior year at the University of Washington, I purchased my first car—an emerald green 1952 two-door Mercury with white leather seats. Boy, was I in hog heaven. That acquisition marked and hastened the engagement with my wife-to-be, and Anna and I were married the day after graduation. Immediately following the completion of my summer job assignments, we rented a Nation-Wide trailer and lugged the bulk of our belongings three thousand miles across the states to Cambridge, Massachusetts, to begin my master's program in nuclear engineering at MIT.

After I finished my nine months of coursework and started the ensuing summer to wrap up my thesis, we decided we would return to the West Coast to complete my doctorate. Our first child had just been born, and we knew his grandparents would want to see him. Hence, I decided on the University of California at Berkeley to complete my PhD, and we made plans to head back across the states in late August.

But the '52 Merc two-door just wasn't big enough to carry all the extras that any young family requires. A crib, a stroller, extra blankets, toys, and diapers all had to go. So we reluctantly decided to sell my pride and joy just as soon as we could find a good used station wagon. After leaving many used-car lots with the disappointment of finding too big a gap between our desires and our wallet, we ran across an ad for a 1956 Nash Rambler station wagon that was in our price range. We drove to the advertised address, and to our delight the white wagon seemed fine. We asked the gentleman presenting the car if we could take it for a short spin. He politely refused—using the excuse that his insurance would not allow anyone else to drive it. But he took us for a short ride, and it seemed to pass the test.

He then assured us that the car was in mint condition. In fact, he insisted it was a one-owner car now in the hands of a dear old lady who drove it only on Sundays to take flowers from Boston to a New Hampshire cemetery to place on the grave of her recently departed husband. Having pity on his elderly friend, who now needed the money more than she needed the car, he agreed to sell it for her. And we fell for it!

We should have wondered why a sweet little old lady would have a special radio antenna affixed squarely in the center of the luggage rack. We also should have questioned how or why she would have convinced this fast talker, who admitted he was in the mobile radio business, to do her bidding. But the price was right, and it had the cargo room needed for our long trip back to the West. So we forked over the cash, and the gentleman strode away—with an unusually extra bounce in his step.

It was only then that I looked over my shoulder to admire Anna driving our "new" wagon on ahead of me. I nearly collapsed when all I could see was a pillar of smoke wheezing from the tailpipe so thick that our new mechanical marvel seemed to completely disappear into an abyss. I couldn't believe how gullible we were. The old wagon got us to California, but our pit stops along the way usually resulted in the purchase of as much engine oil as gasoline. I never found out whether it was just a badly battered engine or if the pistons were simply devoid of rings.

I didn't learn until many years later that one of the earliest applications of radioisotopes in industry was to measure automotive engine wear. By the use of protons or neutrons to irradiate the surface of the part under investigation (such as a ring or a gear), that portion of the metal can be made radioactive. Hence, during operation any wear on that part results in some radioactive material being deposited in the oil lubrication stream. Oil is then readily analyzed to accurately determine the degree of loss of the metal (engine wear). Furthermore, such wear can be analyzed while the engine is operating, without the need for dismantling and reassembly. This is illustrated in figure 62. The savings in both time and dollars to rapidly test new materials should be readily apparent. A savings in time of 1000 percent (a factor of ten) over conventional methods is now commonplace. It is largely because of this ability to fine-tune the tolerances of moving parts that modern engines are so much superior to the engines of my youth.

A relatively new technique has been developed for studying engine components at the high-flux nuclear reactor located at the Institut Laue Langevin (ILL) in Grenoble, France. Neutrons are being used to perform measurements of strain in a variety of engineering materials and components. For example, Volkswagen has been using the ILL neutron-strain scanner to validate calculations on crankshafts whose parts have to resist vibrations generated at high speeds.[1]

As we noted in the previous chapter, radiation is used extensively to control the thickness of rolled metal as it comes through the dies of a modern steel mill. Since most cars and trucks still require substantial amounts of rolled metal in the making of the body of the vehicle, we can appropriately attribute the quality of that shiny fender and hood to the friendly atom. We also noted in the previous chapter that the quality of the iron or steel used in the frame benefits directly from radiation measuring techniques. Furthermore, any flaws in the welds are often exposed by the radiography process and can then be repaired.

Large amounts and different types of glass are used in manufacturing a modern road vehicle—all the way from windshields to headlights. The quality of that glass is assured because of the careful moisture control at the Corning Glass Works (or a competing factory) afforded by neutron probes positioned at strategic locations.

Finally, most of us are very appreciative of the extended tire wear that is now available. As I again reflect on my youth, getting twenty thousand miles of wear on a set of tires was quite an accomplishment. Today, we would not consider pur-

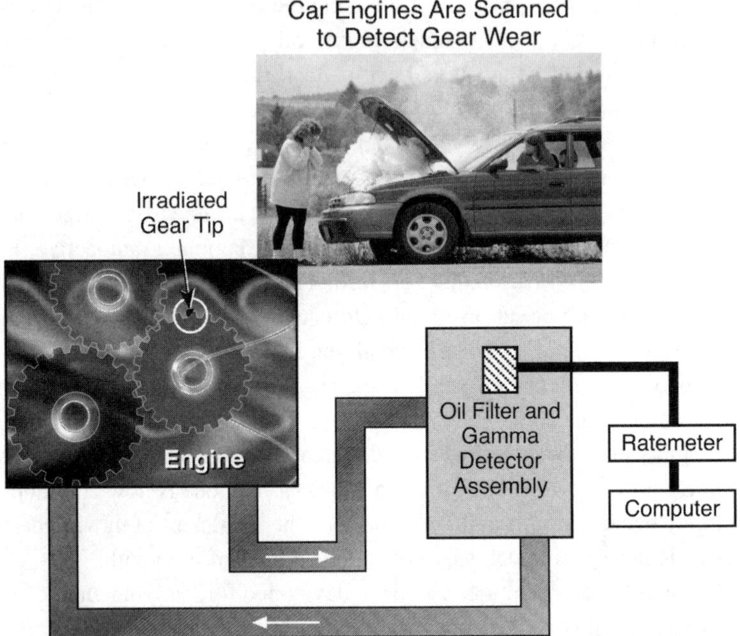

Figure 62. Using radiation to determine the wear on a gear tooth of an operating engine.

chasing tires that did not have at least a forty thousand-mile warranty, and many options are available for twice those expectations. Much of the credit for the improvement in tread wear for car and truck tires is that the rubber is now often vulcanized by radiation, as opposed to the old process of using sulfur. The changes in molecular structure in the rubber induced by the proper exposure to gamma radiation provide both a relatively inexpensive and a superior method to improve the safety and serviceability of automotive tires.

Radiation is also the silent servant behind the scenes in most assembly lines where robots do the bulk of the work. These robots have to be stringently controlled, so radioisotope gauges are often used to produce the precision automation required.

AIRPLANES

It is difficult to imagine how twenty-first-century life could be accommodated without the convenience of air travel. We might put quotes around the word "convenience." This mode of transportation has become a bit more cumbersome since the terrorist attacks of 9/11. But I am penning these words in a hotel room in France, following the conclusion of a three-day workshop relating to the legacy of President Eisenhower's Atoms for Peace speech to the United Nations on December 8, 1953. It was a grueling trip, including a missed connection in Amsterdam, lost luggage, and arrival in a small town southwest of Paris at 12:30 AM, only to learn that the advertised twenty-four-hour taxi service did not exist to get me from the train station to the hotel some three miles away. It was quite an evening.

Nevertheless, I could not possibly have made the trip at all without the availability of an airplane that could transport me from the West Coast of the United States to Atlanta and then across the Atlantic to London and beyond. Given the hectic pace of modern life, the time required traversing the states in a train and then taking a ship across the ocean would have been unthinkable. Air travel has transformed modern life, and there is no turning back.

Since I have traveled well over a million miles on one airline carrier alone, I am acutely aware (and most appreciative) of the safety that has been built into modern aircraft. Whereas air crashes still do occur, the reality is that air travel is by far the safest mode of transportation (as measured by the accident rate per mile). Such a safety record seems counterintuitive, given the speed of such vehicles and the obvious concerns should some type of major failure occur while at thirty thousand feet.

Certainly there are significant safety factors built into the design of such aircraft, knowing that storms and other acts of nature can bounce these airfoils around. But logic dictates that there are practical limits to the thickness of materials that can be used in making the wings and the body, lest the airplane be so heavy that it will never leave the ground. Hence, it has long been a substantial comfort to me to know that the critical welds on those aircraft are routinely inspected by radiography. All commercially licensed aircraft have regularly scheduled inspections in which either neutron or gamma irradiation can be employed to inspect crucial areas such as the welds that attach the wings to the body. Any flaws revealed by this system (illustrated schematically in figure 63) result in immediate attention for appropriate repair. A popular source of neutrons for accomplishing this task is californium-252, a spontaneous neutron emitter that does not occur in nature but is made in nuclear reactors for such purposes. Neutrons are often needed for military aircraft that employ composite construction techniques, whereas gamma sources work well for the metallic construction normally employed for commercial aircraft.

As in the case of the automotive industry, airplanes use substantial amounts

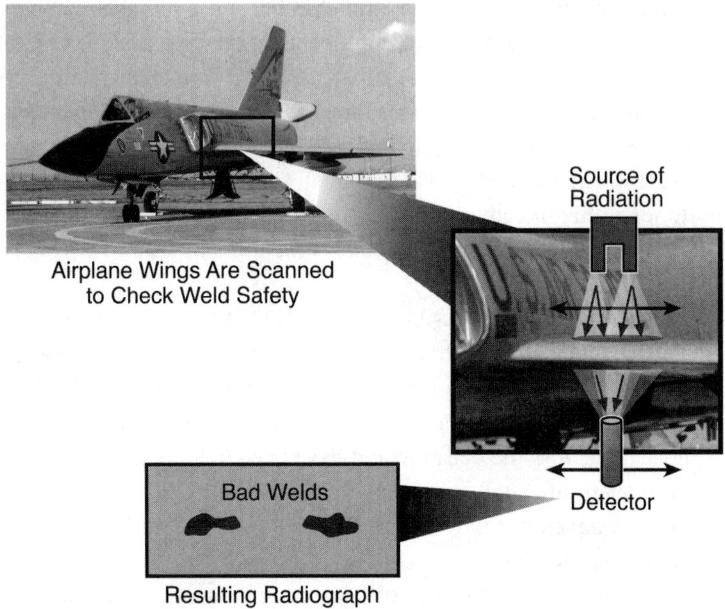

Figure 63. Radiation used to assure the integrity of welds in aircraft.

of rolled metal and glass in their overall construction. Here again, radiation gauges are routinely employed in the manufacture of these materials.

Another unique application of radiation is in providing the source of energy for airport runway lights. Given the possibility of a blackout at airports, one of the special safety features is using tritium to invoke the luminescent property of radiation (illustrated in figure 24 in chapter 3) to assure a twenty-four-hour reliable light source—independent of weather or other confounding factors. Krypton-85 is also used for this purpose. Clearly, radiation provides a major contribution to the safety of modern air travel.

It might be noted, however, that some airline pilots and service personnel have expressed concern about radiation—but from a very different perspective. They recognize that frequent flying at altitudes of twenty thousand to forty thousand feet exposes them to more radiation than if they were on the ground. This is due to less atmospheric shielding of cosmic rays. Indeed, a nonstop flight from Los Angles to New York exposes those in the airplane to approximately 3 mrems of cosmic radiation. Since the annual radiation exposure of an average American is around 360 mrems, one would have to take about a dozen transcontinental flights per year for this to add up to more than 10 percent of his or her annual naturally occurring radiation dose. But professional airline employees spend considerably more time at high altitudes than the average airline traveler. Whereas it is certainly true that professional airline personnel are exposed to more radiation than the normal traveler, it is highly unlikely that any of them would be exposed to a dangerous dose. As we noted earlier, people living in Denver routinely receive about twice the dose of the average American (because of the higher elevation), and there is absolutely no correlation of adverse health effects of such individuals. In fact, the cancer rate of Colorado residents is substantially *below* the national rate. Hence, this expressed concern by airline personnel is almost certainly of little actual substance.

Trains

Many rail systems around the world are electrified, although this is still somewhat uncommon in the United States. Since nuclear energy produces about 17 percent of the total electricity used globally, nuclear power is a significant energy source for trains. As mentioned above, I am writing these words from Saclay, a small village just southwest of Paris in France. Since about 78 percent of the electricity in France is generated though nuclear power, and most of the major rail systems are electrified in that nation, nuclear power is the primary source of energy for the French rail system.

Essentially all of the underground transit systems in the United States are electrified—to avoid the obvious problems of air pollution associated with fossil-fueled propulsion devices. Hence, these transit systems derive their energy from nuclear power in direct proportion to the percentage of the electricity provided locally from nuclear reactors. In the Chicago area, that would be well over 60 percent.

One of the more-unusual applications of radiation to the rail industry worldwide is the measurement of stresses in the rails themselves. A number of serious train accidents occur every year, including fatal derailments that can be attributed to rail fractures resulting from rolling-contact fatigue cracking. Samples of these rail failures have been examined at the European Synchrotron Radiation Facility, located in Grenoble, France, in which 60 KeV x-rays have been used to measure the stress patterns (figure 64).[2] Based on this data, railway engineers have been able to determine the most appropriate practices to minimize the probability of rail failures via proper maintenance and replacement schedules for safe and economic operation.

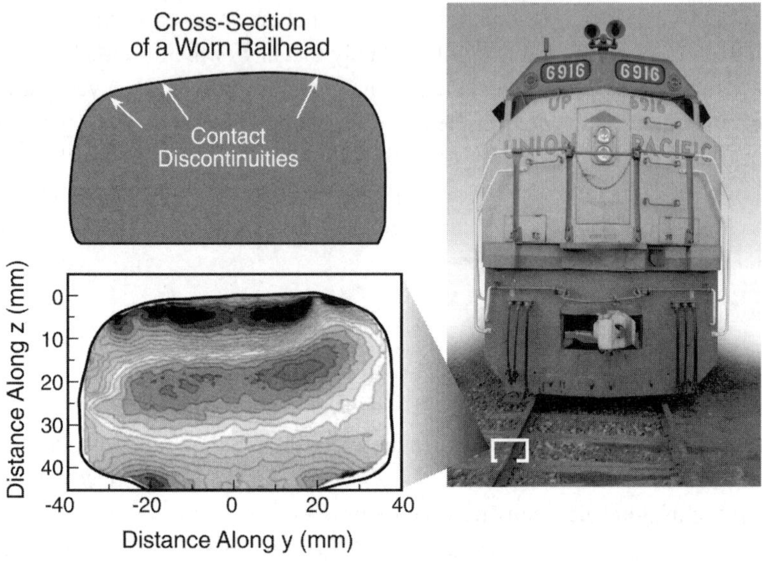

Figure 64. Determining stresses in railroad rails.

SHIPS

We know that nuclear fission was first employed in atomic bombs, which were used to end World War II. It was only later that this same basic energy source was harnessed to heat water in nuclear reactors to produce the steam needed to rotate turbine generators and make electricity. But there was an intermediate step. As an immediate aftermath of World War II, and spurred by the beginning of the Cold War, scientists and engineers recognized that a nuclear reactor might be an ideal source of power for a submarine. Since a nuclear reactor does not require oxygen for combustion (as does a gasoline or diesel-fired engine), a nuclear-powered submarine might be able to stay submerged for a very long time. This, combined with a power plant capable of running for months without the need for new fuel, provided enormous incentive to develop nuclear-powered submarines.

Largely through the leadership of Vice Admiral Hyman Rickover, the US nuclear navy was born. Through sheer tenacity, along with an incredible mind and an unswerving demand for excellence, Rickover was able to deliver the first nuclear submarine—the *Nautilus*—in 1954 (just fourteen years after Enrico Fermi first achieved a sustained chain reaction with his "pile" in Chicago). The *Nautilus* went sixty-two thousand miles on its first reactor core load of fuel, and it traveled ninety-three thousand miles on its second fuel load.[3] This is the equivalent of nearly four complete trips around the globe. Such an achievement was followed by many improvements in design—expanding this monumental success to an entire fleet of nuclear-powered submarines, now numbering about seventy in the United States alone. Many of the modern nuclear subs are able to stay submerged for years and can travel several tens of thousands of miles before they are required to surface. They likewise can travel over thirty miles per hour while fully submerged. Furthermore, the latest US navy submarines need no refueling over their lifetime, which is greater than thirty years.

As indicated above, it was the Naval Nuclear Propulsion Program, with its success in nuclear ships and the pressurized-water reactor power-generating plant at Shippingport, Pennsylvania, that gave rise to the design and construction of the central-station nuclear power plants that generate approximately 17 percent of all the electricity used in the world today. The success of the pressurized-water reactors (PWRs) built to power the nuclear submarines provided the confidence needed by major vendors (such as Westinghouse and General Electric) to dedicate the private funding necessary to start building nuclear reactors for the commercial production of electricity.

Following the success of the nuclear submarine program, the next major application to ships of radiation science was the building of atomic-powered aircraft carriers. The first and most famous of these ships is the USS *Enterprise*. This floating football field is powered by eight nuclear reactors, with two reactors driving a propeller shaft—each generating some seventy thousand horsepower.[4] Given this extraordinary amount of power, the *Enterprise* is able to move at thirty-four knots (nearly forty miles per hour). The success of the *Enterprise* gave rise to the commissioning and building of several more nuclear-powered aircraft carriers. The recently commissioned USS *Ronald Reagan* aircraft carrier is the tenth nuclear-powered carrier built and launched by the United States.

An enormous amount of power is required to service a successful aircraft carrier. In addition to propelling and powering "floating cities" of some five thousand persons, the reactors must supply power to launch scores of aircraft off the flight deck in an amazingly short distance—from a parking position to airborne speeds—via the steam-piston catapult system. Furthermore, since the nuclear reactors do not need to carry any fuel other than that residing in the reactor core, considerable space is available to carry additional aircraft fuel as well as fuel for fossil fuel–driven escort vessels. Although three Bainbridge class nuclear-powered guided-missile destroyers and one nuclear-powered cruiser (USS *Long Beach*) were built, all of the current carrier surface-escort ships are powered by fossil fuel. Hence, the ability of a nuclear aircraft carrier to store liquid fuel that it does not need for its own propulsion provides for substantial flexibility for both extended air operations and in refueling capability for its escort ships during long voyages.

The above vessels are, of course, dedicated to military use. Whereas we noted that the development of this technology was directly utilized to jump-start the central-station nuclear power plants for the commercial generation of electricity, this same technology can be used to build ships to transport commercial cargo. To date, however, only one such US vessel has been built; namely, the *Savannah*.[5] This six hundred-foot vessel has a nuclear power plant that produces twenty-two thousand shaft horsepower, and it can cruise at a speed of twenty-one knots (approximately twenty-five miles per hour). It was designed to carry about sixty passengers and a crew exceeding a hundred, along with a cargo of over seven hundred thousand cubic feet. Its nuclear reactor core was designed to allow three continuous years of operation prior to the need for refueling. The initial fuel load included about 660 pounds of fissile uranium-235. For purposes of comparison, a comparably sized oil-fired ship would consume six million times this weight of oil. This may give us a clue regarding the potential for nuclear power in propelling our large commercial ships of the future.

Russia, which has also employed nuclear power for its submarines, has also utilized nuclear power for several of its icebreakers. A total of nine huge nuclear-powered icebreakers have been in service in that nation to carve winter shipping paths through the frozen Arctic Ocean waters. Some of these icebreakers have been in service for several decades.

ADVANCED FUEL SYSTEMS

Perhaps the most important potential contribution of radiation to the transportation sector is eventually to eliminate the need for oil as the fuel of choice. The transportation sector currently accounts for about one-third of all energy use in the United States. Such a strong dependency upon oil is commonplace throughout the world today. Given the two wars in Iraq within the past decade, few Americans need to be reminded of the price that we are paying for oil derived from that general region. We simply must find an alternative form of energy to power our ships, trains, cars, and airplanes.

President George W. Bush announced in his 2003 State of the Union Address that the United States would initiate an aggressive program to develop fuel cells to provide this energy source. He expressed his hope that by the time babies born at the time of his address were old enough to drive, their vehicles would be powered by hydrogen. Fuel cells, which convert chemical energy directly into electricity, derive their energy by combusting hydrogen and oxygen, leaving only ordinary water as the combustion product.

Indeed, the so-called hydrogen era has enormous appeal. If it can be perfected in a timely manner and at a competitive cost, the dream expressed by President Bush could become a reality. Not only would it replace the prodigious amount of oil currently used for such purposes (thus avoiding the political problems associated with sources of petroleum), but it would also be an environmental masterpiece since it is pollutant free when hydrogen and oxygen are combusted to provide end-use power. Many environmental groups have rushed to the forefront of this new possibility.

But what is the source of the hydrogen?

Currently, the United States manufactures hydrogen from hydrocarbons such as methane. This is accomplished by a "gas-reforming" process, which consists of passing hot steam through beds of hydrocarbons. The problem with this process is that it releases vast amounts of carbon into the atmosphere in the form of CO_2—the gas primarily responsible for global climate change. Hence, this process is not compatible with environmental stewardship. Furthermore, it continues to use up our finite supply of hydrocarbons. Some have advocated solar power as the primary energy source for making hydrogen, but in order to have anywhere near the energy needed to replace one-third of our current energy consumption patterns, entire states would literally need to be covered with solar collectors or wind turbines. Solar energy will likely contribute, but it cannot do the job alone.

Accordingly, nuclear power appears ideally suited as the primary long-term energy source to fuel the hydrogen era. The principal obstacle, at present, is the development of nuclear reactors that can operate at high enough temperatures to achieve the high efficiencies desired in the hydrogen-production process. The method under primary development at present is a sulfur-iodine process that converts water to hydrogen and oxygen, but it needs temperatures upward of 1,800 degrees Fahrenheit (about 1000 °C) to operate efficiently. This requirement is the impetus for current research on advanced nuclear reactors. The challenge is to design a coolant and materials system that can achieve coolant outlet temperatures of this magnitude. Current commercial nuclear reactors are capable only of attaining coolant outlet temperatures of approximately 650 °F. Hence, there are now several efforts underway among the nuclear reactor design community to develop new coolant systems (other than water), along with advanced materials that can attain the elevated temperatures needed for efficient hydrogen production.

Some members of the nuclear community argue that current nuclear reactors could service this growing market through a somewhat inefficient but practical electrolysis process that can be made to work at much lower temperatures—temperatures currently attainable by today's commercial reactors. In order to achieve an economical system, they argue that the reactors would produce electricity during daytime hours (when the demand for electricity is high) and then produce hydrogen at night (when the electrical demand is much lower). This might be practical in some locations because nuclear reactors run most efficiently if kept at constant power—rather than "throttling down" to serve the lower nighttime electrical load demands. Thus, the value of electricity during evening hours is relatively low. This might allow hydrogen to be generated economically even if the conversion process from water to hydrogen and oxygen is relatively inefficient. Another avenue is to search for new hydrogen-generation systems that might allow hydrogen to be efficiently produced at temperatures well below 1,000 °C. Such methods have been identified, but only time will tell whether these approaches will be practical.

Indeed, the possibilities are enormous. At present, nuclear reactors are used mostly to produce electricity (although many in middle Europe and elsewhere are also being used to heat buildings). In the future, the number of uses in the transportation sector will multiply. The potential for nuclear power plants to produce much of the energy now generated by burning oil has monumental implications.

9. Space Exploration

One of the advantages of becoming a "senior citizen" is to have the luxury of observing significant changes over a span of several decades. Such changes are especially spectacular in the world of technology.

I shall never forget my startled first impressions of hearing the "beep," "beep," "beep," of the signals from *Sputnik*, the first man-made satellite surprisingly launched into Earth orbit by the Soviet Union on October 8, 1957. How could such a feat be possible? It hit me like a thunderbolt.

I was a freshman at Centralia Junior College, a small two-year college in southwestern Washington close to the family farm where I grew up. When news hit that this astronomical accomplishment had occurred, I could hardly wait for evening to come—hoping that I could catch a glimpse of reflected sunlight from this awesome tiny ball hurtling through space. I can still recall precisely the spot where I stood in the science building parking lot with my neck craned toward the heavens.

Fast-forward forty-four years. I once again found myself staring into space with a similar sense of exhilaration. It was an evening just days before the kickoff of our Women in Discovery event at Texas A&M University. In particular, we had appealed to the Johnson Space Center for Col. Eileen Collins to come to our university and address a large assembly of high school students as part of our Discovery program. Colonel Collins was the first female astronaut to command a US space mission.

On this memorable night, she was bringing the *Discovery* space shuttle home. Her trajectory from California to touchdown at Cape Canaveral in Florida brought her right over College Station, Texas. The spacecraft at our point was so low as she settled in for a landing that I could almost see the writing on the ship with my naked eye. Wow! From a tiny, unmanned satellite in 1957 to a fully equipped spacecraft with a crew of five in 2001. Flashbacks of the technology that went into that modern spacecraft, which earlier had allowed space pioneers to go to the moon and back, all raced through my mind. I was psyched. My pleasure in actually meeting Eileen was exceeded only by the excitement that she instilled in the hundreds of high school students as she inspired desire in another generation to follow in her footsteps. Indeed, the space age is here. It represents a marvelous combination of intriguing technology and undaunted human courage.

Given this astounding technological triumph, it should come as no great surprise that one of the most amazing ingredients of our entry into the space age is once again part of the legacy of Marie Curie—namely, ionizing radiation. Radiation techniques are used in a variety of essential ways in the National Aeronautics and Space Administration (NASA) program, including the production of heat, electricity, and propulsion.

RADIOISOTOPE HEATER UNITS

A major requirement of any satellite or spacecraft launched into space is to have a mechanism to produce sufficient heat to protect instruments and crews from the cold. The cold vacuum of outer space can dip to four hundred degrees Fahrenheit below zero.* For Earth-orbital missions, it is practical to use solar collector panels to generate electricity and then provide the necessary heat via standard resistance heating. However, for deep-space missions, deriving heat from the sun to energize the solar collectors is not feasible. Weight is the major factor. Massive collector panels are required if the heating requirements are substantial. Since the energy from the sun drops inversely proportional to square of the distance from the sun, the solar energy can become very weak at the outer planets. A satellite orbiting Earth is approximately ninety-three million miles from the sun. By the time a space probe reaches Mars (a distance from the sun 1.5 times that of Earth), the intensity of solar radiation has dropped by a factor of over two. If the probe reaches Jupiter (some five hundred million miles from the sun), the solar radiation intensity is only about 4 percent of that near Earth. (We noted in

* The vacuum of outer space has no temperature in and of itself. Hence, the temperature in a spacecraft is determined by the net balance of energy gains and losses. In low Earth orbit, this naturally balances to between zero and minus thirty-two degrees Fahrenheit.

our story in chapter 1 that by the time *Voyager 2* reached Saturn, the solar radiation intensity had dropped to only about 1 percent of that occurring on Earth.) It simply is not practical to construct solar panels of the size necessary to work under such conditions.

It is likewise infeasible to build batteries that could deliver the needed electricity for resistance heating over such long lifetimes. Many of our space probes require years to reach their destinations. Even our heartiest batteries could not hope to hold up for missions of this magnitude. There is simply no practical way to recharge them.

Fuel cells represent another option. They have been successfully used for providing electricity and heat for several flights. However, both hydrogen and oxygen are required as fuel. Hence, space and weight constraints for fuel storage become excessive for long voyages.

However, it is quite possible to use certain radioisotopes for these purposes. Plutonium-238 is an ideal isotope. Pu-238 decays via alpha emission, with no associated gamma rays. Hence, it is exceptionally easy to shield. The alpha particles are stopped within a small fraction of an inch within an enclosing ceramic matrix, causing the ceramic material to heat up. An additional feature is the almost-ideal half-life of Pu-238, namely, eighty-seven years. This allows such a heating device to deliver approximately 95 percent of its original heat at five years after launch. A full 50 percent of its original heat is still being generated eighty-seven years after the beginning of the voyage.

Because of such desired characteristics, Pu-238 has become the preferred power source for the radioisotope heater unit (RHU). It is lightweight, has a long life, delivers a continuous supply of thermal energy, is resistant to environmental extremes such as temperature, and is completely independent of sunlight.

We hasten to add that Pu-238 is *not* a weapons material. In fact, the presence of any Pu-238 in the mix of the higher isotopes of plutonium that is used for bomb material is quite undesirable. Consequently, the only practical use of this isotope of plutonium is to produce a sustained supply of heat—precisely the attribute needed for space flight.

A disadvantage of the RHU is that the power delivered is actually quite small. One unit typically produces only about one watt of power (about the amount consumed by a miniature Christmas tree bulb). However, by placing several of these units together in a space probe, enough heat is produced to protect the instrumentation from the bitter cold of outer space. Each unit is about 1 inch in diameter by 1.3 inches long and weighs 1.4 ounces. Given this small size and weight, it is practical to put several units together to form a cluster. A typical clustered RHU contains eighty-four of these units, yielding a power output of about eighty-four watts.

As illustrated in figure 65, the core of a RHU consists of the fuel pellet (a), which is made of Pu-238 imbedded within a bricklike ceramic matrix. It is designed to break up into large pieces, rather than disperse as dust, in the event of an accidental impact. This design feature makes it difficult to inhale the plutonium (the major radiological health concern for alpha emitters), should the spacecraft boosters explode on the launch pad or should the spacecraft itself reenter Earth's atmosphere under uncontrolled conditions after being placed into orbit. The cladding (b), which provides structure for the fuel, and insulation (c) are designed to protect the fuel pellet from the extreme heat of reentry and from the environment in the case of an accident. Because of the reliability and robustness of these units, RHUs are used in almost all space missions designed to leave Earth orbit. They are basic to successful flights.

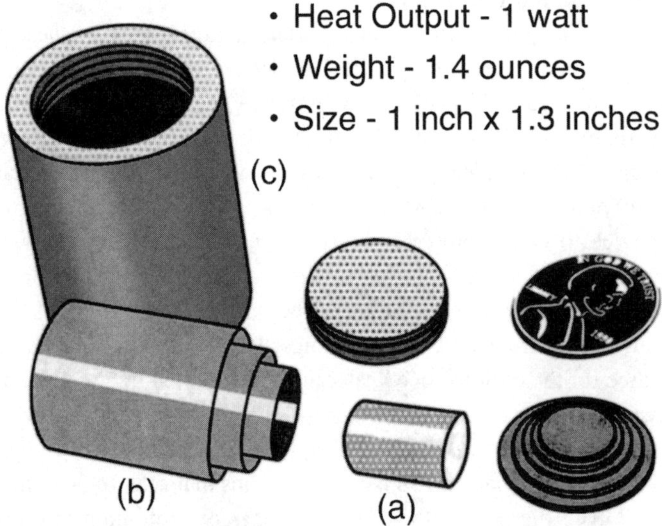

Figure 65. A radioisotope heater unit (RHU).[1]

ELECTRICITY SOURCES

In addition to heating requirements, all spacecraft have a need for a reliable electrical power supply. A variety of instruments must be powered, even for the simplest of unmanned satellites, to provide flight controls, to collect scientific data via photography or other sensing devices crucial to the mission, and to transmit these data back to Earth. Naturally, the demand for electricity grows with the size and the complexity of the mission. Such demands increase from a few watts to several tens of kilowatts for fully manned spacecraft.

The most trustworthy nonsolar source of electricity to meet these requirements is the radioisotope thermoelectric generator (RTG). This device converts heat to electricity via the thermoelectric principle (the Seebeck effect,[2] named for its discovery by Thomas Johann Seebeck in 1821). The device works as follows. If two insulated dissimilar wires (i.e., wires made of different materials, such as nickel and copper) are laid side by side and joined together by junctions at their ends, thus forming a closed circuit, a flow of electrons through these wires can be generated by heating one junction and cooling the other. This principle is utilized in the construction of an RTG by placing a radioactive heat source (the RHU described above) to heat one junction, while cooling the other junction via heat rejection to outer space.

These RTGs are exceptionally robust and reliable because they have no moving parts. The first use of an RTG was in 1961, just four years after the Soviet launch of *Sputnik*. The electrical power provided by this device (a key component of the Space Nuclear Auxiliary Power [SNAP-3] mission) was a mere 2.7 watts. However, it worked as designed and continued to perform for fifteen years after launch. Since that initial launch, RTGs have been used in approximately two dozen US space missions. In addition, astronauts on five Apollo missions left RTG units on the surface of the moon to power the Apollo Lunar Surface Experiment Packages. Advances in power output have allowed newer RTGs to provide all the electrical power required for the Pioneer, Voyager, Galileo, and Ulysses missions.

Polonium-210 was used for the very early space probes in the SNAP-3 series, but was soon abandoned in favor of Pu-238 because Po-210 has a half-life of only 138 days, thus considerably restricting the useful life of the RTG. Pu-238 has proven to be the ideal heat source.

The general purpose heat source (GPHS) is the radioactive fuel package generally used in RTGs for recent and currently planned space missions. These GPHSs use Pu-238, just as earlier-model RTGs did, but the unique fuel containment system is designed to further improve safety in the case of an accident. Each GPHS/RTG contains eighteen modules, as illustrated in figure 66.

The fuel pellets are made of hard, ceramic plutonium-238 oxide that will not dissolve in water. As noted earlier, the pellets are highly resistant to vaporizing or fracturing into breathable particles following an impact on hard surfaces. Each of the eighteen modules contains four fuel pellets, totaling seventy-two pellets per GPHS RTG. Iridium is used to encapsulate each fuel pellet. It is an elastic material that tends to stretch or flatten instead of ripping open should the GPHS module strike the ground at a high speed. A high-strength graphite cylinder surrounds each pair of fuel pellets. This "graphite impact shell" is designed to limit damage to the iridium fuel capsules from free fall or explosion fragments. Finally, an "aeroshell" encloses a pair of graphite impact shells. This outer shield

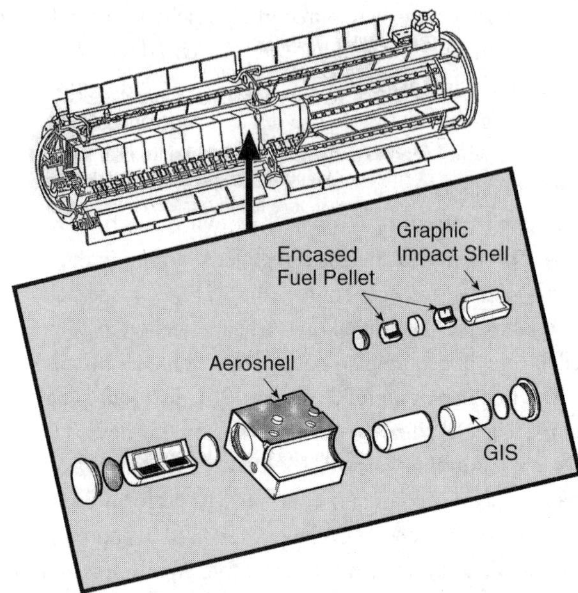

Figure 66. The general purpose heat source (GPHS) used for modern space flights.[3]

is designed to protect the internal components against the heat of reentry into Earth's atmosphere in the event of an accident.

This elaborate defense-in-depth design effort is to assure the safety of these radiation heat sources under all conceivable accident conditions.[4] Rigorous testing is conducted to assure the integrity of the radioactive capsules. Tests include direct exposure to solid propellant fires, arc-jet blast furnaces, impact at 120 miles per hour (approximate top speed for an aeroshell falling to Earth), long-term immersion in seawater, exposure to flying shrapnel, and impact with large fragments (such as steel plates). These tests have demonstrated that the RTGs are extremely rugged and capable of meeting the design objective to prevent or minimize any fuel release.

SUMMARY OF KEY SPACE MISSIONS

Table 2 contains a listing[5] of key US space launches wherein RTGs were used to provide essential power for the mission. A brief explanation of some of the most prominent deep-space missions, beyond the Apollo moon landing campaigns, is included below.

Table 2. Key US space launches where RTGs provided essential power

Power Source	Spacecraft	Mission Type	Launch Date	Status
SNAP-3B7	*Transit 4A*	Navigational	June 29, 1961	RTG operated for 15 years. Satellite now shut down but operational.
SNAP-3B8	*Transit 4B*	Navigational	Nov. 15, 1961	RTG operated for 9 years. Satellite operated periodically after 1962 high-altitude test. Last reported signal in 1971.
SNAP-9A	*Transit 5-BN-1*	Navigational	Sept. 28, 1963	RTG operated as planned. Non-RTG electrical problems on satellite caused satellite to fall after 9 months.
SNAP-9A	*Transit 5-BN-2*	Navigational	Dec. 5, 1963	RTG operational for over 6 years. Satellite lost ability to navigate after 1.5 years.
SNAP-9A	*Transit 5-BN-3*	Navigational	April 21, 1964	Mission was aborted because of launch vehicle failure. RTG burned up on reentry as designed.
SNAP 19B2	*Nimbus-B-1*	Meteorological	May 18, 1968	Mission was aborted because of range safety destruct. RTG heat sources recovered and recycled.
SNAP 19B3	*Nimbus III*	Meteorological	April 14, 1969	RTGs operated for over 2.5 years.
ALRH	*Apollo 11*	Lunar surface	July 14, 1969	Radioisotope heater units for seismic experimental package. Station was shut down August 3, 1969.
SNAP-27	*Apollo 12*	Lunar surface	Nov. 14, 1969	RTG operated for about 8 years until station was shut down.
SNAP-27	*Apollo 13*	Lunar surface	April 11, 1970	Mission aborted on the way to the moon. RTG reentered Earth's atmosphere and landed in South Pacific Ocean. No radiation was released.
SNAP-27	*Apollo 14*	Lunar surface	Jan. 31, 1971	RTG operated for over 6.5 years until station was shut down.
SNAP-27	*Apollo 15*	Lunar surface	July 26, 1971	RTG operated for over 6 years until station was shut down.

Table 2. Key US space launches where RTGs provided essential power (*continued*)

Power Source	Spacecraft	Mission Type	Launch Date	Status
SNAP-19	*Pioneer 10*	Planetary	March 2, 1972	RTG-powered spacecraft successfully operated to Jupiter and is now beyond orbit of Pluto. The last signal was sent to Earth in 2003.
SNAP-27	*Apollo 16*	Lunar surface	April 16, 1972	RTG operated for about 5.5 years until station was shut down.
Transit-RTG	"*Transit*" (*Triad-01-1X*)	Navigational	Sept. 2, 1972	RTG still operating.
SNAP-27	*Apollo 17*	Lunar surface	Dec. 7, 1972	RTG operated for almost 5 years until station was shut down.
SNAP-19	*Pioneer 11*	Planetary	April 5, 1973	RTGs still operating. Spacecraft successfully operated to Jupiter, Saturn, and beyond.
SNAP-19	*Viking 1*	Mars surface	Aug. 20, 1975	RTGs operated for over 6 years until lander was shut down.
SNAP-19	*Viking 2*	Mars surface	Sept. 9, 1975	RTGs operated for over 4 years until relay link was lost.
MHW-RTG	*LES 8**	Communications	March 14, 1976	RTGs still operating.
MHW-RTG	*Voyager 2*	Planetary	Aug. 20, 1977	RTGs still operating. Spacecraft successfully operated to Jupiter, Saturn, Uranus, Neptune, and beyond.
MHW-RTG	*Voyager 1*	Planetary	Sept. 5 1977	RTGs still operating. Spacecraft successfully operated to Jupiter, Saturn, and beyond.
GPHS-RTG	*Galileo*	Planetary	Oct. 18, 1989	RTGs still operating. Spacecraft en route to Jupiter.
GPHS-RTG	*Ulysses*	Planetary/Solar	Oct. 6, 1990	RTG still operating. Spacecraft in route to solar polar flyby.

The Pioneer Missions

The *Pioneer 10* and *Pioneer 11* missions blazed the space trail into the outer planets and beyond. Launched in 1972, *Pioneer 10* and its RTG survived the asteroid belt and the intense radiation field around Jupiter, continuing to perform experiments perfectly. It then continued its journey and finally left the solar system—the first man-made object ever to do so. *Pioneer 11*, launched a year later in 1973, flew by Jupiter with a performance even better than *Pioneer 10*, and then went farther for a close encounter with Saturn. *Pioneer 10* is reported[6] to have sent its last signal back in the spring of 2003, some thirty-one years after its launch—all credited to the continuous power provided by its RTGs. At 7.6 billion miles from Earth, the last signal traveling at the speed of light took eleven hours and twenty minutes to arrive back home.

The Voyager Missions

Voyager 1 and *Voyager 2* were launched within a few weeks of each other in 1977. Whereas they were programmed to use different routes, both spacecraft reached Jupiter and Saturn on their way to the ends of our solar system. They likewise observed close up eleven of Saturn's moons. *Voyager 2*, as we will recall from chapter 1, is the one that sent back spectacular photographs of Saturn's famous rings. It was that spacecraft that then continued on to Uranus and Neptune before disappearing into interstellar space. The RTGs powered versatile and complex instruments, including computers and communications equipment that made it possible for both *Voyagers* to transmit 115,200 bits of data per second back to Earth from Jupiter.

A recent news highlight in *Nature*[7] reported that *Voyager 1* may have ventured across a turbulent boundary near the edge of our solar system, where supersonic "winds" of charged particles from the sun collide with matter from interstellar space. No spacecraft has ever before come even close to this boundary, known as termination shock. Now over twenty-six years and eight billion miles from Earth, the durable *Voyager 1* is believed to have spent about six months traversing this termination shock region, covering a distance of more than four hundred million miles. In addition to data showing a braking solar wind, which reduced the speed of the spacecraft, data from *Voyager 1* indicated a hundredfold increase in the number of charged particles detected in this zone. Again, none of this knowledge could have been gained without the radioisotope power sources—still operating at about 80 percent of full capacity.

The Ulysses Mission

The *Ulysses* mission was a joint enterprise of the European Space Agency and NASA, with the objective of orbiting the sun around its north and south poles (rather than the plane in which Earth orbits). Scientists wanted to study the magnetic fields of streams of particles emanating from the sun in positions never before investigated.

Given the mission objective, we might wonder why it was necessary to use radioisotope power, rather than solar energy. The answer is that the closest that *Ulysses* ever got to the sun was when it left Earth. In order to get *Ulysses* into the desired polar orbital position around the sun, it was necessary to first send it to Jupiter and use the enormous gravity of that huge planet to push the spacecraft out of its ecliptic plane into a polar orbit of the sun. The solar density at Jupiter is only 4 percent of that here on Earth. Hence, a solar panel system large enough to intercept this weak solar energy would add 1,200 pounds—nearly doubling the weight of the spacecraft. By comparison, the RTG weighed only 124 pounds (nearly a factor of ten less). There was simply no rocket booster in existence that could have provided the thrust needed for such a weight over this distance. *Ulysses* was lifted into space in 1990 by the space shuttle *Discovery*. Only a single RTG was used, and it provided all the power required aboard the spacecraft for instruments and other equipment.

The Galileo Mission

Galileo was launched into space aboard the space shuttle *Atlantis* in October 1989. Its eight-year mission was to perform numerous studies of our solar system's largest planet, Jupiter, and its four major moons. Scientists have an intense interest in Jupiter since, unlike Earth and the other planets, Jupiter has retained much of its original composition. Hence, a detailed understanding of this planet may considerably enhance our understanding of the origin of the solar system itself.

Available launch vehicles were not powerful enough to send *Galileo* on a direct path to Jupiter. Hence, a complex route was chosen that first took advantage of the gravitational pull of Venus and then Earth (which required two orbits around the sun), with a final arrival at Jupiter in 1994. *Galileo* consisted of two distinct spacecraft pieces: an orbiter and a separate atmospheric probe named Huygens. About a half year before reaching Jupiter, the Huygens probe departed the orbiter and entered the planet's atmosphere. During a short and frenzied period of sixty to seventy-five minutes, its instruments took atmospheric meas-

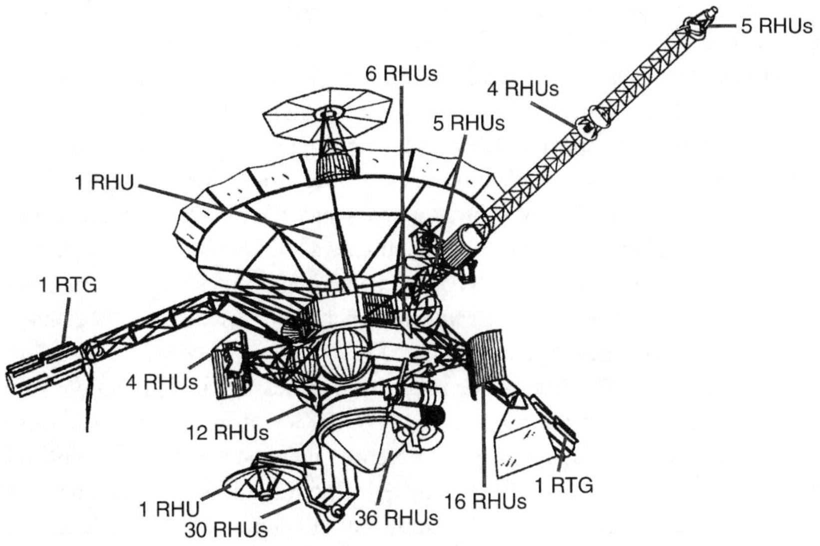

Figure 67. The use of radioisotope heating and power units aboard the *Galileo* spacecraft.[8]

urements and relayed the data to the orbiter for transmission to Earth. The probe then hurtled toward Jupiter's surface until the increasing temperature and pressure of Jupiter's atmosphere crushed and eventually vaporized it.

The orbiter continued to make ten orbits around Jupiter over the next couple of years, obtaining measurements of Jupiter's magnetic atmosphere and observing its four major moons: Io, Europa, Ganymede, and Callisto. None of this could have been accomplished without the RHU and RTG power sources. Figure 67 illustrates the locations and quantities of these sources aboard the spacecraft. It had a total of 120 RHUs and 2 RTGs.

The Cassini Mission

The Cassini program is an international cooperative effort of NASA, the European Space Agency, the Italian Space Agency, and several separate European academic and industrial contributors. The goal of the *Cassini* mission is to perform close-up studies of Saturn, its rings, moons, and magnetic environment. Titan, which is Saturn's largest moon, is of particular interest because of atmospheric and possibly surface characteristics somewhat similar to our early Earth. A key feature of the *Cassini* mission is to parachute an instrument-laden probe to Titan's surface to directly sample the atmosphere and provide our first view of its surface. The global interest and financing arrangements for this mission make it one of the most ambitious and challenging interplanetary explorations ever mounted. Academic and industrial partners in thirty-three states in the United States plus sixteen European nations are providing direct support. To accumulate and transmit the voluminous data expected to be derived from this mission, *Cassini* has been equipped with exceptionally elaborate instrumentation and computational capabilities. *Cassini* has aboard a data system ten times more efficient than the *Galileo* spacecraft, but at less than one-third the mass and volume. Power for this equipment is provided by RTGs, which are supplying a total of 630 watts of electricity. In addition, approximately 117 RHUs are being used to regulate temperatures on the spacecraft and its probe.[9] *Cassini* was launched on October 15, 1997, and it is expected to reach Saturn by about June 2004.[10]

Whereas the RTGs have made impressive and essential contributions to the space program, the overall thermodynamic efficiency of such devices is under 10 percent. That means that less than 10 percent of the total heat energy supplied by the radioisotope sources is converted into electricity. At least two other technologies are being pursued to increase the overall efficiency. The desire is to either reduce the weight and amount of radioisotope material required for a given power demand or increase the power available from the quantities currently used.

One approach is the dynamic isotope power system (DIPS). The word *dynamic* is applied because this system employs moving parts. The Stirling engine is one such design. It still uses Pu-238 as the heat source, but helium is utilized to drive a piston back and forth (much like in an ordinary car engine) to run an alternator that generates electricity. The radioisotope on the hot side of the engine heats the helium, and the helium is then cooled (via heat rejection to outer space) on the other side. This design offers the possibility of attaining a thermodynamic efficiency substantially higher than the RTG. Ironically, while penning this portion of the present chapter, our local newspaper featured a front-page story announcing a $23 million, seven-year contract received by Stirling Technology Company (located in Kennewick, Washington) to fund development of this DIPS engine for NASA. One prototype module was reported to be operating flawlessly in the company's laboratory for over nine years. The NASA design goal is to have such devices run for fourteen years without maintenance or breakdowns.

Another approach under active investigation is the alkaline metal thermal to electric conversion (AMTEC) technique. This device converts thermal energy into electricity using liquid metal ions (charged particles). A thermophotovoltaic (TPV) converter changes infrared radiation emitted by a hot surface into electricity. Investigators using this approach hope to attain thermodynamic efficiencies of 20 to 30 percent, which would be a threefold increase over the RTGs. The basic heating source would again be Pu-238.

NUCLEAR ELECTRIC POWER

As space exploration missions increase in size and complexity, the demands for electricity become increasingly larger. Radioactive sources are clearly of enormous benefit in providing heat and electricity for space probes, but they are limited to a few kilowatts in any practical sense. Furthermore, there is no way to change power levels to meet fluctuating load demands. Nuclear reactors, on the other hand, can be operated at essentially any power desired—up to several megawatts using today's technology. And the power levels of a nuclear reactor can be raised or lowered as desired.

Figure 68 illustrates the approximate ranges of electrical power and the usable lifetimes for the various energy sources available for space travel. We note that batteries are practical for only low-power, short-duration applications. Chemical dynamic engines can deliver enormous power for short periods of time, but these power levels cannot be maintained for a very long time. Solar energy can be employed for much longer durations if the solar density is sufficiently high, but there is a maximum power level that can practically be achievable (given the limitations of solar energy intensity). Photovoltaic cells can be productively used for low-power demands, but solar concentrators (devices to focus solar energy for thermal-mechanical conversion) would be needed for larger electricity requirements. Fuel cells likewise play a major role, but they need both hydrogen and oxygen supplies for operation. Hence, their primary limitation is the requirement for storage of these necessary fuel ingredients. We note here that radioisotope sources can contribute a very long life, but the total power available is limited. Nuclear reactors, on the other hand, can produce a very high range of power and, properly designed, can also operate for quite long periods of time.

To date, the United States has built and launched only one nuclear reactor into space. That sole event was the SNAP-10A mission, which had a small reactor that produced only about 500 watts (0.5 kW) of electrical power. It operated forty-three days in space before being prematurely shut down because of a mechanical relay failure in the spacecraft. The ground version of this same engine operated successfully for over four hundred days. SNAP-10A is still in orbit, and it is expected to remain there for at least another three thousand years. The Soviet Union placed considerably more emphasis on space nuclear power and launched around thirty nuclear reactors into Earth orbit between 1967 and 1988.

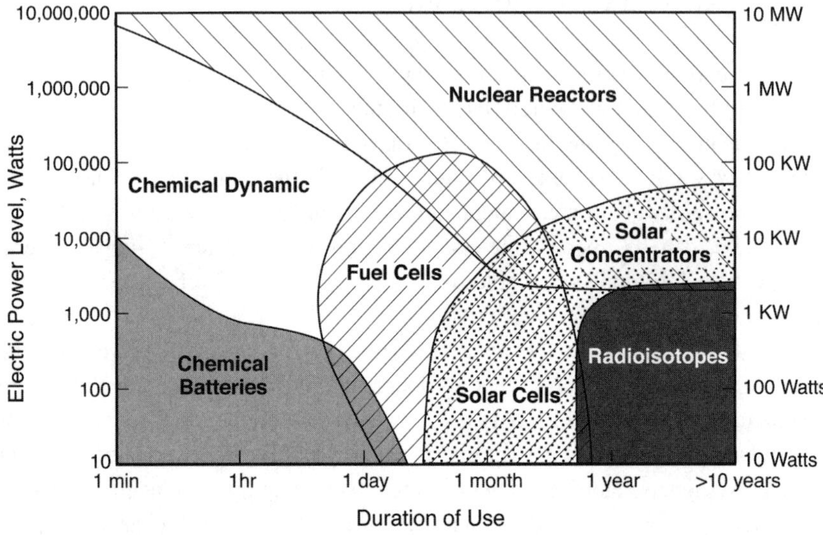

Figure 68. A comparison of energy sources for providing electrical power in space.

NUCLEAR PROPULSION SYSTEMS

The advantages of nuclear reactors for space propulsion are many. First, these systems can provide a high range of power—and do so without the need of oxygen. Nuclear reactor systems are considerably lighter than their chemical counterparts, when both engine and total fuel supply are taken into account, and they may remain usable for up to about fifteen years (much longer than any chemical system).

Given these advantages, NASA has broken its nearly two decades of silence on nuclear fission power and is aggressively pursuing the development of nuclear-powered systems for interplanetary space travel. As James Crocker, vice president of Civil Space at Lockheed Martin Space and Strategic Missiles, aptly said, "99.9 percent of our solar system is unexplorable without nuclear power."[11]

Nuclear power can be utilized for space propulsion in basically two ways. One is to employ a coolant such as hydrogen that also serves as the propellant. This is the nuclear rocket, schematically depicted in figure 69. The second method is to utilize electricity produced by a nuclear reactor to power an ion engine or an arc-jet engine.

A nuclear rocket utilizes the same fuel as employed in commercial central-station nuclear power reactors here on Earth, uranium-235, but in a different chemical form. The reactor is designed so that liquid hydrogen is injected into the reactor core cavity for cooling. The heated hydrogen is then ejected through the rear nozzle

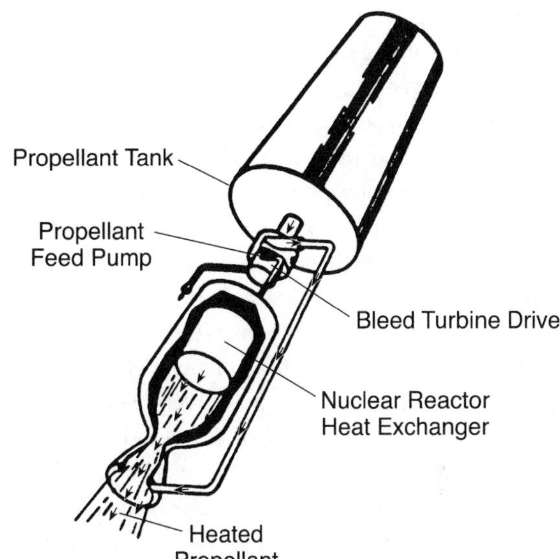

Figure 69. A sketch of a nuclear-powered rocket engine along with its propellant tank of liquid hydrogen.

at a very high velocity. This has been tested many times, and it has been shown to work (although substantial development effort remains to allow these engines to be ready for reliable flight missions). Since hydrogen is the lightest element in the periodic table, it has what is known as the highest specific impulse of any element (i.e., it is inherently the most efficient propellant for rocket performance).

During the early NASA years, a substantial investment of about $1.4 billion was made to develop nuclear-powered rockets. The Nuclear Engine for Rocket Vehicle Applications (NERVA) program[12] resulted in the development of twenty nuclear rockets, which were tested at the Nevada Proving Grounds near Jackass Flats. These reactors demonstrated power levels ranging from 44 MW to 4,000 MW of thermal energy. The upper level is higher than the largest nuclear power reactors that currently provide central-station electricity. Thrusts for these designs were measured to span a range of 50,000 to 250,000 pounds.[13] That compares to maximum chemical booster rocket thrusts of the clustered liquid-fueled Saturn engines at the rear of the space shuttles of 400,000 pounds. The total thrust of the combined Boeing 747 engines at full takeoff power is about 220,000 pounds.

The principal reason that NASA temporarily shelved the development of nuclear rocketry was because of public fears. Some influential public figures expressed fear that a nuclear-powered rocket may blow up on the launch pad and spread contamination. Others were concerned that the development of powerful nuclear rockets for purported civilian use may be a smoke screen for the militarization of space. Yet without such development, any hopes of a "game-changing" approach to space travel is severely thwarted. There is simply no alternative for obtaining the long-term, high-thrust power sources necessary for major new space missions. Hence, the approach at present in reintroducing nuclear power into the space program is to use conventional chemical booster rockets to propel the nuclear rocket package into Earth orbit and not fire up the nuclear rocket until the appropriate time to send the package out of orbit and into the outer reaches of space. This would greatly mitigate any worries over safety concerns of a launch mishap. Still, substantial attention is always given to the safety implications of any launch accident.

The ion engine system utilizes electricity to accelerate ions through the rocket nozzle to provide thrust. This technique is currently being used in some satellites for low-power repositioning maneuvers. Solar power normally provides the electricity for these relatively small devices. The much higher thrusts achievable by the large amounts of electricity that can be generated by a nuclear reactor are still substantially less than those attainable by heating and expanding hydrogen. However low-thrust, long-operating systems are very important for interplanetary travel. The high-thrust engines are used to enable the space vehicle to escape the gravitational pull of the planets, but the ion engines are expected to be sufficient to maintain and enhance vehicle velocity outside of the strong gravitational fields.

The latest large NASA mission is called the Jupiter Icy Moons Orbiter (JIMO), in which a nuclear-powered engine will be developed to provide the power necessary to drive an ion engine suitable for this very long mission. The nuclear reactor portion of the project, which will feature a nuclear reactor designed to produce approximately a hundred kilowatts of electricity, is called Project Prometheus.

Nuclear-generated electricity can also be used to power an arc-jet engine, a relatively futuristic design that promises considerably more thrust than possible with an ion engine.

A MANNED VOYAGE TO MARS

One of the most enjoyable experiences that I had as a college professor was going out to high schools to provide juniors and seniors with a glimpse of the career opportunities that the pioneering work of Marie Curie has provided our new generation. Without question, the topic of most interest to these bright, young students was the lure of space travel. In particular, I usually showed a short film clip of the famed "Vomit Comet."

The Vomit Comet is, in actuality, not a comet at all. Rather, it is a KC-135A, the military tanker version of the Boeing 720 (an earlier version of the famed

Boeing 707). It is flown by NASA pilots to train astronauts and other personnel on techniques necessary to cope with zero gravity. This airplane is normally taken to cruising altitude and then put into a steep climb. Near the top of the climb, the pilot throttles back and arcs the airplane over in a smooth parabolic motion to point the nose down toward Earth. During about a twenty-second interval, the combination of countering the upward momentum and then descending allows the occupants of the aircraft to experience very nearly zero gravity. They actually float inside the shell of the aircraft. Astronauts get crucial training in this environment, because it is the only place on Earth where zero gravity can be simulated. Interestingly enough, this is also the place where space films are often made. It is where Tom Hanks and his crew filmed the movie Apollo 13.

But in addition to being a serious training laboratory for space flight, NASA has graciously opened up this program to special college programs. Students compete heavily from all over the nation for this trip of a lifetime. It does not come without a lot of hard work, however, since in order to qualify, the students must design and build an experiment acceptable to NASA that provides for the conducting of serious scientific work in zero gravity. If their project is selected, the students must then go to the Johnson Space Center in Houston, Texas, and undergo physiological training. Their reward is a series of approximately three dozen "cycles" in this specially equipped aircraft. After each downward plunge by the aircraft, the pilot then points the nose back upward and provides full throttle to bring the airborne laboratory back up for another parabolic rollover. During the bottom phase of the cycle, the occupants experience a harrowing 2Gs (twice normal gravitational forces). They are instructed to lie on their backs during this phase of the flight. Each microgravity flight lasts for two to three hours while the airplane is making thirty to forty parabolic maneuvers. It is because of the constant fluctuation between the skin-stretching 2Gs and the heavenly floatations that has, on more than one occasion, caused the stomachs of some students to rebel. Hence, the name Vomit Comet.

But what a thrill it must be. I have talked to several of our students who were fortunate enough to make the trip, and they all refer to it as the adventure of a lifetime. Short of becoming an astronaut, it is the closest thing to space travel that we mortals could ever experience.

Ever since Yuri Gagarin was launched into Earth orbit by the USSR aboard the *Vostok* spacecraft on April 12, 1961, the real lure of the space age is manned missions. Alan Shepard ignited the American spirit on May 5, 1961, when he climbed aboard the *Mercury Freedom 7* capsule to become the first American launched into space. Putting real people into space vehicles was why the Apollo moon-landing program was so exciting. To actually have a human step foot on the moon and then return to Earth was a feat that justifiably captured headlines throughout the world. It was that program that provided serious thought to be given to the idea that planet Earth may not necessarily be our only home for future generations.

It is for this reason that the vision of sending a manned mission to Mars has captivated so many people. Whereas there is no expectation that Mars might actually be habitable, such a mission would mark a major milestone in humankind's ability to explore another planet and return to tell about it. One of the major difficulties in planning such a mission, however, is to find an energy source with sufficient thrust and longevity to allow the trip to be made in a reasonable amount of time.

The use of chemical rockets would make such a voyage arduously long and hazardous. Current estimates are that a manned flight to Mars using conventional chemical boosters would take at least six months to get there and another six months to get back. Given travel uncertainties, plus time to actually explore the Red Planet, the round-trip could take up to two or three years.[14] Nuclear propulsion devices could cut the trip to as little as two months, thus allowing the entire round-trip to be approximately a half year in duration. Another significant advantage of nuclear propulsion is that it would open up many more launch opportunities (it would not constrain planners to seek particular periods of time when the planets were optimally lined up). Furthermore, there would be sufficient power in a nuclear engine to provide all the other power needed on the flight (heat for the crew and instruments, refrigeration as needed, and flight maneuverability).

An additional concern for the long flight required by chemical propulsion is astronaut muscle atrophy, owing to the near zero gravity during the flight to Mars. Astronauts experience this problem on much shorter flights. A Mars mission would hold dubious promise if it were to land a half dozen physically disabled astronauts who were expected to actively carry out significant exploration upon arrival. One way to solve this problem is to have sufficient power aboard the spacecraft to energize thrusters capable of placing it into a controlled end-over-end spin.[15] With the crew habitat module at one end of the spacecraft and the nuclear-rocket core affixed to the other end, the craft would swing around a center of mass located near the inside end of the engine stage. This would provide a "downward" centrifugal force, pressing the crew toward the outside end of the habitat module. If, on the way toward Mars, the craft could be forced to tumble at about four rotations per minute, this would create a gravity about the same as they would experience on Mars (about one-third that on Earth).

Other International Space Endeavors

As noted earlier, the United States has launched only one nuclear-powered spacecraft into Earth orbit, whereas the former Soviet Union placed some thirty-three nuclear reactors into space between December 1967 and March 1988. The reactors launched by the USSR were of the RORSAT (Radar Ocean Reconnaissance Satellite) series. They were small reactors, cooled by liquid sodium-potassium (NaK), and producing approximately 2 kW of electricity. They used a thermoelectric converter, which allowed them to achieve a thermodynamic efficiency of only 2 to 4 percent. The Soviets elected to use this form of nuclear power, rather than RTGs, because their purpose was to power radar equipment to keep track of major naval maneuvers of NATO and the US Navy.[16] Table 3 contains a listing[17] of the USSR launches, which featured the predominate use of nuclear reactors for their electrical power supply.

Toward the end of this program, the Soviets realized that they needed to place satellites into higher orbit and that more electricity would be required to power their radar equipment. Hence, they designed and launched some early TOPAZ-type reactors. These reactors, using thermoionic energy converters, were capable of providing 5 to 6 kW of electricity, and they were also shown to attain higher thermodynamic efficiencies—as high as 10 percent.

Until recently, the United States and the former Soviet Union were the only nations to launch manned space vehicles. However, China recently entered the picture with the successful launching of the manned *Shenzhou* 5 on October 15, 2003. The largest international collaboration in space has been associated with the International Space Station. Led by the United States, this station is drawing upon the scientific and technological resources of sixteen nations, including Canada, Japan, Russia, Brazil, and eleven nations of the European Space Agency. Electrical power for the space station is being supplied by nearly an acre of solar panels.

Table 3. USSR space launches featuring the use of nuclear reactors

Date	Spacecraft	Power Source	Mean Altitude	Status/Lifetime	Notes
3 Sep 65	*Cosmos 84*	RTG	1,500km	—	—
18 Sep 65	*Cosmos 90*	RTG	1,500km	—	—
27 Dec 67	*Cosmos 198*	Reactor	920km	1 Day	—
22 Mar 68	*Cosmos 209*	Reactor	905km	1 Day	—
25 Jan 69	*RORSAT* Launch Failure?	—	—	—	—
23 Sep 69	*Cosmos 300*	RTG	Reentered	—	Lunar Probe?
22 Oct 69	*Cosmos 305*	RTG	Reentered	—	Lunar Probe?
3 Oct 70	*Cosmos 367*	Reactor	970km	1 Day	—
1 Apr 71	*Cosmos 402*	Reactor	990km	1 Day	—
25 Dec 71	*Cosmos 469*	Reactor	980km	9 Days	—
21 Aug 72	*Cosmos 516*	Reactor	975km	32 Days	—
25 Apr 73	*RORSAT* Launch Failure?	—	—	—	—
27 Dec 73	*Cosmos 626*	Reactor	945km	45 Days	—
15 May 74	*Cosmos 651*	Reactor	920km	71 Days	—
17 May 74	*Cosmos 654*	Reactor	965km	74 Days	—
2 Apr 75	*Cosmos 723*	Reactor	930km	43 Days	—
7 Apr 75	*Cosmos 724*	Reactor	900km	65 Days	—
12 Dec 75	*Cosmos 785*	Reactor	955km	1 Day	—
17 Oct 76	*Cosmos 860*	Reactor	960km	24 Days	—
21 Oct 76	*Cosmos 861*	Reactor	960km	60 Days	—
16 Sep 77	*Cosmos 952*	Reactor	950km	21 Days	—
18 Sep 77	*Cosmos 954*	Reactor	Reentered	~43 Days	Canada Impact
29 Apr 80	*Cosmos 1176*	Reactor	920km	134 Days	—
5 Mar 81	*Cosmos 1249*	Reactor	940km	105 Days	—
21 Apr 81	*Cosmos 1266*	Reactor	930km	8 Days	—
24 Aug 81	*Cosmos 1299*	Reactor	945km	12 Days	—
14 May 82	*Cosmos 1365*	Reactor	930km	135 Days	—
1 Jun 82	*Cosmos 1372*	Reactor	945km	70 Days	—
30 Aug 82	*Cosmos 1402*	Reactor	Reentered	120 Days	South Atlantic
2 Oct 82	*Cosmos 1412*	Reactor	945km	39 Days	—
29 Jun 84	*Cosmos 1579*	Reactor	945km	39 Days	—
31 Oct 84	*Cosmos 1607*	Reactor	950km	93 Days	—
1 Aug 85	*Cosmos 1670*	Reactor	950km	83 Days	—
23 Aug 85	*Cosmos 1677*	Reactor	940km	60 Days	—
21 Mar 86	*Cosmos 1736*	Reactor	950km	92 Days	—
20 Aug 86	*Cosmos 1771*	Reactor	950km	56 Days	—
1 Feb 87	*Cosmos 1818*	Reactor	800km	~6 Months	Topaz 1
18 Jun 87	*Cosmos 1860*	Reactor	950km	40 Days	—
10 Jul 87	*Cosmos 1867*	Reactor	800km	~1 Year	Topaz 1
12 Dec 87	*Cosmos 1900*	Reactor	720km	~124 Days	Malfunction
14 Mar 88	*Cosmos 1932*	Reactor	965km	66 Days	Last RORSAT

Radiation Health Effects of Space Travel

Whereas the above discussion reveals the clear contributions that radiation has made in our efforts to conquer space, there is a downside. Astronauts face many hazards when they volunteer to be strapped into a space capsule and prepare to be propelled off a launch pad. Among them are an explosion during liftoff (or anytime later in the flight), a complete dependence on only the oxygen and food that they carry with them, and of course potential failures of any equipment that is needed to bring them safely back to Earth.

But there is another danger, and that is ionizing radiation. We noted in chapter 2 that cosmic radiation is very much a part of our universe. We are shielded from much of this on Earth because of our protective atmospheric layer. But once out of this layer, the only protection the astronauts have is the skin of their capsule. Under most conditions this is quite sufficient, since cosmic radiation is actually quite sparse. However, during certain "sunspot" conditions, considerable radiation is emitted by the sun, and this has the potential to render physical harm to the astronauts. One way to cope with these radiation showers is to maneuver the capsule such that the heat shield (that familiar surface that is covered with heat-resistant tiles) is placed between them and the oncoming radiation.

Still, the greatest protection is to have a vehicle with sufficient power to get them to and from their target in minimum time. Here again, nuclear propulsion is by far the best solution—particularly for flights lasting several months or more. Therefore, if we are serious about really conquering the "last frontier," radiation technology has come to our aid at just the right time.

10. Terrorism, Crime, and Public Safety

The world has never been a safe place in which to live. As far back as history has been recorded, we know that staying alive has been a constant struggle. Wars, floods, famines, pestilence, . . . you name it. In some ways it is a wonder that the human species has been able to survive, much less thrive.

Some will say that life is even more risky today. In reading the morning headlines, we see a daily threat of death and destruction in some part of our globe. We are constantly reminded of earthquakes, forest fires, floods, volcanic eruptions, and hurricanes, along with a new generation of epidemics such as AIDS and SARS. And if natural disasters and epidemics were not enough, we now have the constant threat of terrorism—either on a local or a massive scale. In numerous quarters, there seems to be no shortage of reasons for open hostility, so war also remains a constant threat.

Then, of course, there are the injuries and deaths associated with accidents and disease. Some would argue that modern technology has led to a much more hazardous world than in previous generations. Almost every day the media proclaim a new link with cancer associated with some element of technology. If we could have only lived in the "good old days" when life was easier and when the world was safer, they seem to say.

To be sure, we are not immortal. Yet despite the much faster pace of life that modern technology has brought us, the facts show us that life is constantly becoming safer. Since the Industrial Revolution in the late 1800s, we are enjoying a substantially longer life span. Life expectancy in the Western world was only about fifty years when Marie Curie was doing her pioneering work. A century later, life expectancy is in excess of seventy-five years.

This is not to downplay the hazards that occur in everyday life. Rather, it is to suggest that modern technology has brought us a plethora of comforts that our parents and grandparents could hardly have dreamed of. It is instructive, therefore, to reflect on some of the ways that modern technology has been developed to protect us against new threats and, in particular, how radiation has been employed to enable such protection.

PERSONAL SAFETY

Like many others, I have a special passion for the theater. But any time a throng of people becomes wedged into cramped quarters, there is always the potential for fire or other safety threats that provide an impetus for prompt evacuation. In such situations, it is comforting to know that the omnipresent exit sign will always stay illuminated, even in a blackened theater, because it is powered by radiation. By employing the twenty-four-hour reliability of the luminescence property of radiation, that sign can be relied upon to operate under any and all circumstances. The glass tube spelling "exit" is painted with a luminescent material and is filled with tritium gas. The constantly emanating electrons from the tritium impact this special paint and cause the emission of light visible to the human eye. Such signs are now located in most public buildings in the Western world and have undoubtedly saved many lives when buildings needed to be evacuated during power failures and other hazardous conditions.

Figure 70. A typical exit sign powered by radiation.

Other illumination applications include airport runway lighting and luminous dials. Most modern airports line their crucial runways with lights powered with the radioisotopes tritium or krypton-85 so that proper visual illumination is guaranteed—even under the most unusual circumstances (including earthquakes). Luminous dials were once a most valued commodity, particularly for aircraft instruments prior to the time of proper lighting in the cockpit. However, most of this need has faded as better illumination provided by electricity is now widely available.

Perhaps the most noticeable application of radioisotopes to modern industrial safety is the common household smoke detector. This device contains a small amount of a special radioisotope (often americium-241, although plutonium-239 has been used in numerous smoke alarms in Russia) that constantly emits alpha particles into an ion chamber. Hence, during normal atmospheric conditions, a small but steady electric ion current is actively recorded. However, if a fire occurs, smoke particles enter the chamber and reduce the ion current. The change in current triggers the smoke alarm. These compact smoke alarm devices have saved untold lives as well as valuable property throughout the world.

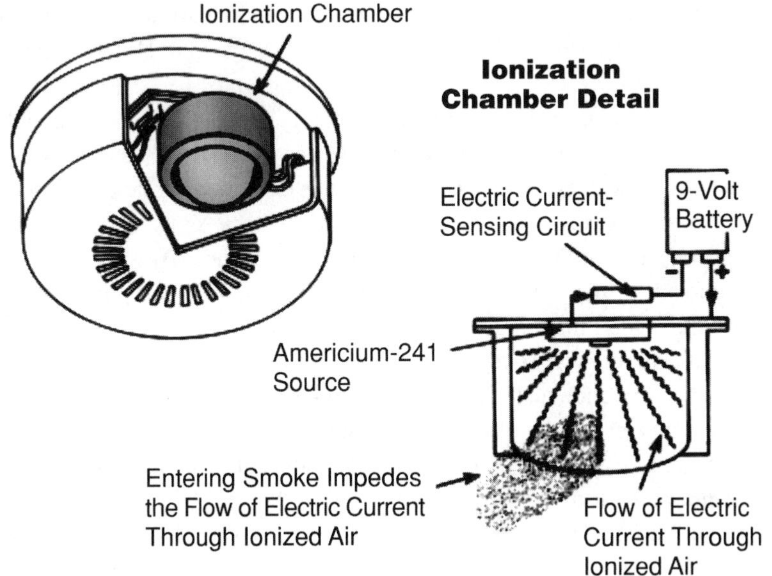

Figure 71. The common household smoke detector.

There are numerous other areas in which radiation contributes to making life safer. One rather unique application is in the paper-duplicating industry. As most of us know, static electricity can build up when we make a few paper copies on a home duplicating machine. This can become a major safety hazard when expanded to the commercial scale. Hence, radiation from sources such as polonium-210 can and is being used to nullify the static electrical charge buildup. By installing devices that emit charged particles, the positive charge in these operations can be conveniently nullified. This provides a powerful safety measure in the paper-duplicating industry and for those who photocopy large numbers of papers.

Although not directly related to safety, the process mentioned above is often used in scientific laboratories to remove the effects of static electricity that would otherwise disturb sensitive balancing measurements. Radium sources have been used for these applications, since its decay chain contains several alpha and beta emitters that can effectively neutralize static electricity.

FIGHTING CRIME

Crime has undoubtedly been a distasteful part of humanity as long as history itself. And with the advancements in technology, crime has become increasingly more sophisticated. Consequently, it is necessary to employ new techniques in order to both thwart criminal activity and provide ways to track down and convict the offenders.

As might be expected, the activation property of radiation is a strong ally to aid in criminal investigations. Being able to trace the paths of guns, knives, and personal effects such as gloves and shoes can be exceptionally helpful in solving difficult criminal cases. Even trace amounts of "tell-tale signs" can be analyzed by using radiation techniques to help bring about justice. By sampling only a small amount of dirt on the shoes of a suspect in an appropriate radiation detector, the precise ingredients of the sample can often be matched to other soil samples to determine where the suspect previously has been.

Neutron activation analysis (NAA) was used as early as the 1950s by criminologists. This process involves placing a sample of evidence found at the crime scene into a nuclear reactor and irradiating it with neutrons. The neutrons collide with components of trace elements contained in the sample, causing them to be transmuted into other isotopes that subsequently decay and emit gamma radiation of a characteristic energy level. This allows accurate identification of many constituents in the sample. The amazing thing about this procedure is that it will work for a small sample of hair. As many as a dozen or so different elements can be identified in a single strand of hair.

To the best of my knowledge, the first case to utilize neutron activation analysis in a major crime was that of the 1958 murder of a sixteen-year-old Canadian girl named Gaetane Bouchard.[1] Her former boyfriend, John Vollman, lived just across the border in Maine and was seen with her just before she was discovered dead. Flakes of paint from the place where they had been together were matched to his car. An additional piece of evidence was the victim's color of lipstick, which was found on candy in the glove compartment of the suspect's car. But it was the strands of hair found clasped in the victim's hand that convinced the jury to convict Mr. Vollman of the murder. These hair strands were matched to Vollman via the neutron activation procedure, wherein the ratio of sulfur to phosphorus contained in this hair sample was closer to his ratio than hers.

One of the great mysteries surrounding the aftermath of the 1815 battle of Waterloo, after Napoleon went into exile, was what caused his death. Although controversy still rages, many scholars now believe that the evidence points to arsenic poisoning. The basis for this conclusion is that strands of his hair, salvaged from the time of his death, were tested in the 1960s by using neutron activation analysis. The diagnosis provided convincing evidence of toxic arsenic traces.[2] How the arsenic got into Napoleon's body is still debated, but some have concluded that it was deliberately administered in his wine.[3] If so, it provides one with a greater respect for the level of trust the great kings of the world had in their cupbearers!

Having grown up watching Old West movies, I recall several instances where a gunslinger was brought to justice when the sheriff was able to detect gunpowder on the suspect's hands. The procedure was fairly straightforward. It involved melting a glob of paraffin in a pot and then painting it onto the fingers and hands of the suspect. The wax was then peeled off and treated with chemicals that react to gunpowder traces. If the chemicals turned up positive, the sheriff had his man. The only problem is that such chemicals can often produce the same results when reacting with bleach, fertilizer, or urine. This often diminishes the ability of this procedure to prove guilt. In a recent article in *Time* magazine,[4] Barry Fischer, director of the Los Angeles sheriff department's forensics laboratory, outlined some of the newer and more precise procedures. One of the drawbacks of the modern procedures, however, was that "you don't get to see the terror on the people's faces when you pour hot paraffin on their hands. I think it encouraged some people to confess."

But aside from the stimulus of "encouraged confessions" via the hot paraffin technique, modern procedures are far more accurate. Today, most forensic labs searching for trace amounts of gunpowder rely on scanning electron microscopes. This is accomplished by touching a bit of tape to the hands of the suspect and then placing the tape under the microscope and hitting it with a stream of electrons. Since the elements in gunpowder will then respond to electron bombardment by emitting distinctive x-rays, even minute amounts of gunpowder can be easily and precisely detected.

One modern technique that is being used more frequently to bring criminals to justice is DNA analysis. Exceptionally minute quantities of "residuals" associated with suspected persons can be used to provide positive identification. By matching residuals at the scene of the crime (such as blood stains, saliva, semen, or hair) with the suspect, prosecutors have powerful evidence in their arsenal. Such analyses have been used in famous cases and in other circumstances, such as in the O. J. Simpson murder trial, the President Clinton incident with Monica Lewinsky, and the positive identification of the deaths of Saddam Hussein's eldest sons. What may not be known to many people is that radiation often plays a key role in the analysis.

"DNA fingerprinting" has gained considerable popularity and glamour in the recent past because we now know that every person on Earth has his or her own unique variation of the DNA molecule—the only exception being identical twins. This affords forensic experts the possibility of providing a positive linking of crime scene trace DNA residue to only one person in the universe. However, the processes involved in "DNA fingerprinting" are quite complex.[5] Hence, DNA analyses have sometimes been discarded in courts of law because of the lack of precision in the sample collection or in the measurement techniques. Also, since DNA analyses are time consuming and expensive to perform, other "fingerprinting" techniques are being developed for more frequent application.

To get a perspective on the complexity of the sorting process necessary to identify the DNA of an individual, we'll take a little "time out" to sketch the biological makeup of this very important molecule. DNA stands for deoxyribonucleic acid, a molecule located in the nucleus of a cell that is part sugar (*ribose* sugar) that has lost a critical oxygen atom (thus *deoxy*). It is the basic building block of life. Under the power of a special microscope, these molecules look like a rubber ladder twisted at one end (the double helix). These DNA "ladders" are very long, literally about a yard in length, yet they have to fit in a human cell only a minute fraction of an inch in diameter. As Dr. John Medina notes in his excellent book on the genetic makeup of the human body,[6] it is like taking thirty miles of gold thread and stuffing it into a cherry pit. But despite this compactness, it is highly ordered in a manner capable of providing instructions to the complex biological process that we call life.

Each cell in the human body contains twenty-three pairs of chromosomes, and it is the DNA molecule that makes up the chemical structure of these chromosomes. Chromosomes typically hold about three thousand functioning genes, with each gene being either dominant or recessive. Each gene has two possible configurations, and it is the way these genes are configured that codifies the traits of an individual. If all of this encoded information were written with a standard typewriter, this would create an encyclopedia some three hundred volumes in length with five hundred pages per volume.[7] Beyond this encoded library, upward of 98 percent of the DNA contained in the chromosomes is essentially meaningless (i.e., not containing any particular traits). Hence, this portion is often called "junk DNA."

The key to all of the above is that the DNA (whether in the genes or in the junk part) is composed of small biochemical units called *nucleotides*. DNA fingerprinting is based upon getting a copy of these unique variations in the arrangement of nucleotides. The nucleotides arrange themselves in pairs of four unique types (abbreviated as A, T, C, and G) to form the rungs of the DNA double helix–shaped ladder.

It should now be obvious that the number of variations possible in such a massive molecule makes us all unique. But we should also realize that given this size and degree of complexity, it is impossible for the whole genetic code of an individual to be examined. Hence, the techniques used for DNA fingerprinting are based either on the type of genes an individual has or upon the amount of reoccurrence of a certain pattern of nucleotides in his or her genetic material.[8] The former approach to determining a DNA fingerprint is called the polymerase chain reaction (PCR) technique, wherein the gene types of an individual are matched. The latter approach is called the restriction fragment length protocol (RFLP) technique, wherein the repetition of nucleotide patterns within a chromosome is measured to obtain a DNA fingerprint.

Once a sample to be DNA fingerprinted is sent to the laboratory, the DNA must first be removed from the host material by a neutral solvent solution. The solution is then put into a centrifuge and spun at very high speeds until the cells containing DNA are separated. The final step is to place these cells into a solution of enzymes that eat away the cell membranes to isolate the DNA molecules.

The PCR method does not involve radiation techniques in the DNA-matching process. Rather, it is based upon duplicating a single strand of DNA in an exponential fashion, such that twenty replication cycles yields about a million copies of the original DNA. These new copies are placed on a sheet of nylon. Gene probes, which bind to certain genes, are then introduced, and they will attach to specific genes on the DNA strand. A special indicator solution, unique to each gene probe, is then added, and when this indicator locates a probe that is attached to a "target gene," a blue dot appears on the nylon strip. This series of blue dots and blank misses is the "DNA fingerprint." The advantage of the PCR method is that it requires only a very small sample size—perhaps the size of the tip of a pin. Furthermore, it takes only about two or three weeks to complete the procedure. However, the disadvantage of the PCR method is that it is considerably less accurate than the RFLP method discussed below. The probability of error in matching individuals with the PCR sample may be from one in a hundred to one in two thousand. Whereas this might sound quite precise, it may be insufficient to convince a jury "beyond a reasonable doubt." In fact, it was largely due to this possibility of error that the analysis provided by this approach was discarded in the O. J. Simpson trial.[9]

The RFLP method is based on employing an enzyme that can locate a certain string of nucleotides in the DNA molecule. After cutting the DNA into a series of variable lengths, these pieces are put into a special type of gel at one end of a container that has electrical plates affixed to both ends. An electrical current is then passed through the solution in a process known as electrophoresis. The pieces of DNA move through the gel at different rates because of their different sizes. The gel is subsequently removed and bathed in a solution containing four types of radioactive probes, each of which binds to a different nucleotide on the end of the cut-up DNA pieces. A piece of x-ray film is then placed on top of the gel, and a radioactive tag, normally phosphorus-32, is then used to expose the film. The strips of white at different locations on the exposed film indicate how far the different pieces of DNA have moved. This pattern is the classic "DNA fingerprint."

This procedure can take a much longer time to complete than the PCR method, primarily because of the small amount of phosphorus-32 that is tagged to the DNA. It may require up to two weeks to significantly expose a piece of x-ray film. In addition, the film must be separately exposed for each of five or more areas being tested, each taking an additional week. Upon completion of the procedure, the film is often scanned into a computer to obtain precise measurements. All of this could take three or four months. However, the reward is that the accuracy is exceptionally high. It is from five to a thousand times more accurate than the PCR method. In fact, some of the experts at trials have testified that the error in this method, based on radioactive techniques, may be only about one in a hundred million. When a prosecutor tries a case and can show that the chance of the DNA's not being that of the defendant is one in a hundred million, he has very compelling evidence.

It should be noted that in both cases it is essential that the sample being measured contains nucleated cells. This is the case for blood, saliva, semen, or skin. It is also true for hair if the root is attached, but a strand of hair detached from its root cannot be analyzed by this method. However, a new procedure, known as mitochondrial DNA, was recently developed, which is often used for hair, bone, and teeth.[10] The FBI began using this technique in 1996.

To sum up the DNA-fingerprinting approach, this area is getting increasing attention as a way to match the DNA found in substances at the scene of a crime with the unique DNA of a suspect. Whereas very precise matching can be done using the approach based on employing radioisotopes, the process is expensive and time consuming. Hence, nonradioactive techniques are becoming popular for routine investigations. But for situations where preciseness really matters, the RFLP method, which relies on radiation techniques for crucial steps, remains an important forensic tool. An error rate of only one part in a hundred million can free an innocent person or send a murderer to jail.

One of the hottest new tools for solving crimes is called infrared microspectroscopy (IMS). This technique relies on the focusing of an infrared spectrum on a sample and detecting the different frequencies of light that then emanate from it.[11] The problem is that it is very difficult to obtain infrared sources strong enough to get precise measurements, particularly for small samples. However, this technique has been improved considerably by obtaining a very strong infrared spectrum from a synchrotron electron beam accelerator.[12] When electrons are accelerated in a circular accelerator, like that of the Advanced Light Source (ALS) facility at Berkeley,[13] they release light through beam lines that can be focused on a very small sample of suspicious material, causing the yield of other light frequencies, which can be measured by an infrared detector. This information can be used to identify the ingredients of the sample. Whereas the basic physics processes utilized in this approach do not depend upon ionizing radiation, the intensity of the infrared light has been made vastly more powerful by the use of accelerators like the ALS at the Lawrence Berkeley National Laboratory.

The amazing thing about the accelerator approach is how little of the suspicious material is required in order to derive a positive analysis. In one case, colleagues accused a businessman of cocaine abuse—a charge he vehemently denied.[14] But a glass plate was subsequently recovered from the accused businessman's office and closely inspected. Even though it had been meticulously cleaned of all contaminants, a microscopic examination revealed the presence of some shallow scratches. Only a few tens of nanograms* of material were obtained from within the scratches, but it was sufficient to allow the IMS method to demonstrate convincingly the existence of cocaine.

Fighting Terrorism

September 11, 2001, will undoubtedly be recorded in history as one of the most significant days ever in shaping our hopes and dreams for the future. Its events

*A nanogram is one-billionth of a gram

were unlike any other—with the possible exception of the assassination of President Kennedy. Everyone I have ever met who was of a mature age or even younger in 1963 remembers precisely where they were and what they were doing when they heard of the November 22 assassination in Dallas. September 11, 2001, had a similar effect on our consciousness.

I was in Paris when 9/11 occurred. I was attending an international conference dealing with advanced fuel cycles for future nuclear energy systems, and I will never forget staring at the television monitors in stark disbelief. Needless to say, the tenor of the conference changed dramatically. That evening I decided to calm my nerves by taking a long walk along the Champs Elysees. But rather than this stroll providing solace, I could not help looking at the Arc de Triumph and other famous landmarks wondering how a free society could ever prevent a lightning-fast destruction of such masterpieces—much less the deaths of thousands of people. I then reflected back on my presence at the US Embassy in Paris only one day earlier, totally unaware that it was also a terrorist target—but a plot that had been foiled. It causes one to wonder what has become of humanity.

Prior to the 9/11 attacks, airline hijackings had already begun to take their toll, and it was clear that measures were needed in airports to thwart such attempts. Airline passengers had long become accustomed to x-ray devices at check-in counters, where baggage could be rapidly screened for metal or other potentially dangerous objects. One fairly recent development, however, is the application of neutron activation to detect hidden explosives. A technique now employed in some airports is to direct a beam of neutrons on samples taken from passenger luggage. This transmutes a material such as nitrogen (a common ingredient in explosives) into a radioactive isotope that emits characteristic radiation that can be detected. Whereas this provides positive identification of suspicious materials, this procedure is generally not suitable for high-speed throughput of luggage inspections. Hence, most airports currently utilize an ion mobility spectrographic system. In many cases, a beta emitter such as nickel-63 is used to create the ionization required for sample investigations. This is likely the procedure used when airport security personnel take a "swipe" around your personal laptop computer or other carry-on luggage.

Such measures are, of course, not foolproof. They did not prevent the hijackers from boarding airplanes on the morning of September 11 and carrying out their morbid plans. On the other hand, that was before airport security personnel were aware that zealous terrorists existed who were willing to sacrifice their own lives through suicidal missions.

Airport security has been substantially enhanced since that time, as we all know. Radiation devices, which form the core of those new airline screening instruments, continue to play a key role in fighting terrorism. The US government has moved swiftly to be sure that all of the 429 commercial airports within its continental borders have baggage-screening machines.[15] However, many of these machines are still capable of producing only x-ray pictures, which often cannot distinguish between a block of plastic explosives and a wedge of cheese. This places considerably more reliance on the operator than is desirable—and it is why more-advanced detection machines are being designed for implementation.

In some instances, it is necessary for monitors to have the capability of determining the contents of packages that are several feet in depth, such as in the case of cargo inspection at entry ports. C-van type containers aboard ships, trucks, and trains account for approximately 90 percent of the world's traded cargo when measured by value. They account for 95 percent of the US import and export cargo tonnage. Figure 72 illustrates a typical load of C-vans aboard a ship arriving at a major US port. In 2002 some 6.8 million containers flowed in and out of global borders, and this is expected to quadruple in the next twenty years.[16] The sobering aspect of this mammoth movement of goods is that in 2002 only about 2 percent of such cargo was selected for inspection. Consequently, there is a huge effort underway to develop scanners that have the capability of nondestructively penetrating depths of several feet through heavy metal containers and can do so with sufficient accuracy and speed to avoid slowing up the flow of commerce that is so essential for modern living. The properties of radiation make the use of both gamma and neutron sources logical choices for this task.

One system currently being deployed relies on either cobalt-60 or cesium-137 to supply the intense gamma rays needed for this deep penetration.[17] Both of these radioisotopes can be assembled in sufficient concentrations to provide credible results. Neutron sources of this intensity are more difficult and expensive to obtain, but the potential use of neutrons for such screening is very high. This has spurred interest in developing pulsed neutron sources using a variety of techniques. One approach that I became aware of during my stint in Texas was that of developing a very intense burst of neutrons utilizing the collapsing electrical fields of huge capacitor banks to form pulsed plasmas.[18] If approaches such as this could be perfected, and then produced and operated economically, it would likely be possible to perform very sensitive screening procedures with incredible precision and speed at all points of entry into the country. Questions of appropriate shielding to protect drivers and other personnel would, of course, need to be properly addressed.

Another area of concern regarding terrorist activities (or cleanup after a war) is the presence of land mines. Numerous dismemberments and deaths occur in war-torn areas because of such devices. They are triggered by stepping on them

Figure 72. C-vans arriving at a major US port.

(shallow burial) or by being run over with heavy vehicles (medium burial). One approach to sensing these land mines is being pursued at Sandia National Laboratory by using an x-ray fluorescence system.[19] This device works by firing x-rays into the ground and measuring the different types of x-rays that return. X-ray returns that match the energy-spectrum characteristic of explosive materials provide a clear warning of buried land mines. Another approach, developed by the Pacific Northwest National Laboratory, is to use neutrons to detect the abundance of hydrogen that resides either in the plastic casings of such land mines or in the explosive material itself. By using electronic discrimination techniques to filter out the neutrons that bounce back from the surface from those thermalized by the hydrogen contained in the land mines, fairly precise identifications can be made.[20] With millions of buried land mines worldwide, the clearing of these devices to prevent injury and death needs to be accomplished in a fast and safe manner.

It was shortly after the shock of 9/11 that the United States became immobilized with the news that anthrax had been sprinkled on some of the letters entering the domestic mail system. Anthrax-laced letters appeared in the offices of key members of the US Senate, causing complete evacuation of several federal offices for a substantial length of time. As officials scrambled for ways to screen mail to be sure it was free of this deadly "dust," I was contacted as head of the Nuclear Engineering Department at Texas A&M to see if radiation could be used in some fashion to kill anthrax spores. The timing of the inquiry was providential, because at precisely that time a senior official from the Titan Corporation was in town to discuss a new project involving the donation of three state-of-the-art accelerators to the university for a new food-irradiation research center. I asked if he would join some of our faculty that afternoon to discuss the possibility of using similar accelerators to irradiate mail.

I quickly learned that he and his colleagues were already well into these kinds of investigations,[21] and key members of our faculty added assurance that, indeed, the needed "kill rate" of anthrax spores could be achieved by the combined beta ray and x-ray capabilities of these accelerators, and at an acceptable throughput rate for mail to key governmental agencies. Still, I was a bit surprised to learn only about a week or so later that Titan was providing several of these accelerators to the US government for use in irradiating mail going to the nation's capital. It is amazing how rapidly decisions can be made in a time of public crisis.

Whereas there clearly would be logistical difficulties in requiring similar sterilization for the full mail system of any major country, and certainly so for the United States, this system did work for the limited volume of mail needed to keep the wheels of government moving in the United States. Improvements on the system have been subsequently made to refine the process and avoid many of the initial problems, such as causing some of the paper to be slightly discolored.[22] Who, only a few years ago, could have conceived that such radiation devices would become the strong ally of our mail carriers and the public?

But another new threat has come about with the onset of mass terrorism; namely, the possible development and use of a radiation dispersion device (a RDD, or "dirty bomb"). I believe that it is important to give this possibility particular attention. The concern is that terrorists could get access to radioactive material and mix it into a standard high-explosive chemical bomb. Upon detonation, the radioactive material would be dispersed and would frighten a large number of people.

Indeed, this is a very credible possibility. If one considers the great number of places where radioactive material could be obtained (laboratories, hospitals, medical clinics, etc.), there certainly are ways that a dedicated terrorist group could gain access. Yes, security measures are being stepped up in most places, but the plethora of sites makes it essentially impossible to prevent dedicated terrorist groups from obtaining small samples of radioactive material.

The question is how much damage could an RDD really do? This is a very important question. Without considerable preparation for public understanding, my concern is that panic could break out—which is precisely what the terrorist group would want to happen. I feel compelled to say at the outset that it is *possible* for significant harm to be rendered to the public, should such an attack occur. Many of my colleagues scoff at the dangers associated with an RDD. They argue that it would be essentially impossible for a perpetrator to include enough harmful radioisotopes in such a device to impart a significant public health hazard without the instigators themselves being killed in the process of putting the device together. The effect of any hazardous radioactive material embedded in the device would be far more dangerous at close range than it would be after being detonated by conventional chemical explosives and dispersed at great distances. Indeed, this is all true. The sticking point, however, is that we have now learned that some terrorists are so depraved and brainwashed that they have no regard for the value of life—and they are willing to commit suicide to achieve their goals. Since even very high doses of radiation do not cause death for several days, it is possible for a deranged, dedicated group of terrorists to assemble such a device knowing that they would not survive. This type of mentality and sacrificial zeal is a relatively new phenomenon.

Having said this, however, I believe it is even more important to say that it is *highly unlikely* that a significant public health hazard would occur from the detonation of such a device. The reasons are threefold. First, the major hazard of any detonation at close range is the force of the chemical explosion itself. This is clearly evidenced by the terrorist bombings in recent years that have occurred so frequently in Iraq and other parts of the Middle East. Anyone within a few tens of feet from a major conventional chemical explosion is certainly in considerable danger of a violent death. The addition of a large radiation component would be of secondary order. Second, even if a massive amount of hazardous radioactive material could be assembled and placed into an RDD, the danger drops by the square of the distance from the source of detonation. Thus, at a point ten feet from the detonation, the danger would drop by a factor of a hundred. At a hundred feet, it would drop by a factor of ten thousand. If one were a mile away from the detonation (5,280 feet), the hazard would drop by a factor of over twenty-seven million. These are averages, assuming no filtering by the atmosphere or wind direction considerations. Clearly, any radiation concerns become very small at reasonable distances from the point of detonation. Third, I personally believe that the actual dangers of radiation to the human body are considerably less than those conservatively accepted by the present agencies that set the acceptable radiation standards. My reasons for this were outlined in chapter 2.

Hence, my urgent plea is that if terrorists should either threaten the explosion of a "dirty bomb" or actually detonate one, please listen to the instructions of responsible safety officials (e.g., the chief of police or the fire chief). *Do not panic!* There is a very high probability that the actual public health hazard is far less than what the terrorists would like us to believe. Remember, their intent is to create panic, and they believe that the mere mention of the word *radiation* will achieve their goals. It is my fervent hope that we will be sufficiently forewarned not to fall for their attempts at deception.

The Question of the Bomb

Whereas the topic of atomic bombs may or may not seem to fit the theme of the present discussion, the fact is that atomic (and hydrogen) bombs have been developed, and we cannot put that genie back into the bottle. Hence, we have to find ways to live in the world where one giant mishap could bring destruction unheard of in any previous era of human history.

Such weapons would not be possible without the properties of nuclear fission and fusion. Yet it is these same properties that hold such enormous promise for the peaceful use of nuclear energy. The harsh fact is that we cannot ignore the "genie" that has been loosed. The challenge is to hold the destructive force (i.e., bombs) in check while exploiting the positive side to the fullest service to humanity.

In a broader sense, it may be inappropriate to look upon the weaponry side of the atomic age with only distaste. The reality is that the harnessing of atomic power in this form ended World War II with far fewer casualties than would have been the case had conventional warfare continued—though no one could responsibly argue that the dropping of these bombs on Japan was not a tragedy. It could be argued that the nuclear capability held by the West during the previous half century has helped to maintain some form of stability in the world. The mere fact that powerful Western nations such as the United States could employ nuclear warheads formed the centerpiece of the checks and balances crucial to world stability. But whatever view the reader may take on this issue, the fact is that such technology is readily available, and unilateral disarmament (though advocated by some) would almost assuredly lead to disaster.

Given the events of 9/11, there is now interest in radiation technology to fight terrorism. Examples of where radiation technology is essential in this fight were noted earlier, such as to detect metal or explosives at airports and public facilities. But this technology is also essential in sensing clandestine weapons testing, sensing contamination releases, and having the ability to disable weapons of mass destruction. The latter could become even more important in the future. The possibility of a terrorist group flaunting a biological, chemical, or nuclear weapon of mass destruction is almost certain to cause extreme fear among members of the public. Hence, it is important to develop measures to disable such systems (without the disabling system itself being a weapon of mass destruction!). One such measure would be to develop compact systems capable of generating intense bursts of neutrons or other radiation of sufficient intensity to kill biological threats (e.g., anthrax) or otherwise disable the electronic or magnetic firing mechanisms of a nuclear device.

We live in a new world and need to have every technology available in developing effective countermeasures. Radiation technology is certainly one essential tool.

RISK PERSPECTIVES

As we close this chapter on public safety risks, we may need to be reminded that whether we are addressing safety from either a strictly personal perspective or the more general public viewpoint, it is essential to pause and always ask, "relative to what?" Wishful dreaming aside, there is no such thing as absolute safety. Every gulp of air that we breathe and every bite of food that we swallow carries some risk. The reason we do these things, however, is because the risk of not doing so may carry an even greater risk. Simply stated, we recognize the benefits in eating and breathing, and we will likely continue these practices—whether or not we are even cognizant that we are continually making such choices.

One of the persons who has helped me the most in understanding the concept of risks and benefits is Dr. Bernard Cohen, a retired professor of physics at the University of Pittsburgh. Professor Cohen has done extensive research in studying and publishing the risks of everyday living.[23] I was so impressed that I sought and obtained his permission to reformat many of his results and publish them in one of my previous books.[24] For instance, I learned that the biggest hazard to the health of a male is to be unemployed and living in poverty. This is not a joke. Statistics clearly reveal that the life patterns formed by being forced to live in poverty actually rob such a man of thirty-five hundred days of normal life expectancy. Perhaps an even bigger surprise is that the second-biggest hazard for a male is to remain unmarried. An unmarried male, on average, loses three thousand days of normal life expectancy. It turns out that unmarried females also live a shorter life span than their married counterparts, but they lose only sixteen hundred days of normal life expectancy. That's right. We guys need our women more than they need us!

The reason for bringing up such points at this juncture of the book might seem humorous to some, and perhaps totally irrelevant to others. But I do this for a very important reason. The *actual* risks to everyday living may be (and usually are) far different than our *perceptions* of risks. I know that after 9/11, many people are afraid to do things that they once took for granted. They just don't feel the world is a safe place to live in anymore. No group is better aware of this than the airlines and the general travel industry. Yet the actual hazards of terrorist attacks in most parts of the world are far less than the hazards we often inflict upon ourselves, just because of the seemingly unconscious choices that we make in our daily lives.

I was again reminded of this by an article in our local newspaper in the spring of 2003.[25] Garret Condon reported in the *Hartford Courant* (and replicated in other newspapers such as ours) that we need to develop a balanced perspective in times of uncertainty. Citing several scientific studies, Mr. Condon points out that heart disease is a much greater threat than anthrax. He quoted the Web site of the Harvard Center for Risk Analysis as reporting that the annual risk of dying from heart disease is one in 397, whereas it is one in 56,424,800 for anthrax. That's nearly a factor of 150,000. He then quotes Dr. David Katz, associate clinical professor of public health and medicine at the Yale School of Medicine, as noting that obesity kills between three hundred thousand and four hundred thousand Americans per year. That dwarfs any statistics associated with terrorism. Translating the obesity concern to Dr. Cohen's methodology, a person thirty pounds overweight can expect to lose 1,020 days of his or her normal life expectancy.

The point of this section is to remind us to keep a healthy outlook on life. Whereas it is true that crime is still a major concern, war seems to remain inevitable, and the threat of terrorism is likely here to stay for quite some time into the future, our chances of personal injury from this dark side of humanity are actually quite small. We make choices every day that likely have considerably more to do with our personal health and well-being than could ever be forced upon us from the depraved population of our planet. Furthermore, the harnessing of Marie Curie's atom has armed us with a plethora of new tools to continue fighting off those who are so desperate in their living conditions that life no longer holds meaning, promise, or hope. The atom, in conjunction with mutual respect, tolerance, and generosity, may be able to move this desperate situation to a future of hope for all peoples of the world.

11. Arts and Sciences

You are likely aware by now that there is hardly a facet of modern living that does not benefit from radiation in some fashion. This is even true in both the arts and the sciences.

Understanding Our Origins

Once our basic needs of food, clothing, and shelter are satisfied, we humans seem to have an unquenchable thirst for knowing our origin and how the world around us became manifest. A prime example of this curiosity surfaced in my little corner of the world only a few years ago when a major anthropological discovery was made under most unusual circumstances.

On July 28, 1996, two young men were swimming in the Columbia River near Kennewick, Washington, when they accidentally stumbled over a human skull on the bottom of the river about ten feet from shore. They immediately called the police, who contacted the county coroner, who in turn called in Dr. James Chatters, a well-respected anthropologist. Dr. Chatters, noting something unusual about the skeleton, helped police recover much of the complete skeleton—all within a few feet of the original skull location.[1] Thus was the discovery of what has come to be known as "Kennewick Man."

The unique thing about this discovery is that experts have subsequently conducted a scientific dating analysis on a small bone fragment and have made a preliminary determination that Kennewick Man is about eighty-four hundred years old. This revelation has dumbfounded the anthropology community, because this would be one of the oldest and most complete skeletons found in the Northwest.[2] It would bring a whole new perspective on how people started populating the Pacific Northwest. Did they really cross a "land bridge" across the Bering Strait from Asia (presumed to have occurred some four thousand to six thousand years ago)? Kennewick is three thousand miles south of Alaska. How could Kennewick Man have arrived at that location some twenty-five hundred years earlier? Adding a further twist to the dilemma, specialized dating techniques have revealed that people have been in the Americas for at least 11,500 years.[3] Could this person have come from Europe, rather than Asia?

To compound the mystery, considerable controversy emerged from day one regarding who has access to studying Kennewick Man. Archeologists and anthropologists want to study the skeleton in great detail for scientific reasons. On the other hand, native Indian tribes mounted a serious legal campaign to prevent such study, arguing that respect to their early forefathers needs to be honored and they want no part of any disfiguring that might occur if the remains are placed in the hands of scientists. This legal tug-of-war, which may not get resolved shy of the US Supreme Court, has created quite a stir in our little community.

But whether or not the court decides Kennewick Man can be further studied, the question for us at present is what role does radiation play in this controversy? How do we really know the age of this specimen? How can we be certain that Kennewick Man had even older "relatives" in other parts of the Americas?

The answer is the carbon dating process, based on the radioactive decay of carbon-14. Carbon dating was developed just after World War II by a team at the University of Chicago led by Dr. Willard F. Libby.[4] This development, which has allowed us to accurately date a plethora of historical artifacts, has been so monumental that Dr. Libby was awarded the Nobel Prize in Chemistry for his achievement. Dr. Libby knew that natural carbon is mostly carbon-12. It contains six protons and six neutrons in its nucleus, and it constitutes 98.89 percent of all carbon in existence.[5] However, carbon-13 also exists in nature (comprising essentially the remaining 1.11 percent). But of most interest to us is the fact that carbon-14 also exists in nature as a result of cosmic radiation. Neutrons from outer space react with nitrogen in the upper atmosphere to form carbon-14 plus a proton. The amount of carbon-14 is exceptionally small. In fact, for every carbon-14 atom existing in nature, there are 1,000,000,000,000 (one trillion) atoms of carbon-12.

Despite this exceptionally small amount of carbon-14, Dr. Libby and his team recognized that this miniscule quirk of nature could provide a powerful mechanism for dating historical artifacts. The trick was recognizing that there exists a constant ratio of carbon-14 to carbon-12 naturally occurring in our environment at all times. In particular, all living beings (whether plant or animal) have this same ratio in their systems, since they all have a constant intake of carbon while still alive. But when such plants or animals die, there is no more intake of carbon in any form. Carbon-12 is stable (it does not decay), but carbon-14 decays with a half-life of 5,715 years. Consequently, the ratio of carbon-14 to carbon-12 diminishes with time in dead species. This change in the ratio of carbon-14 to carbon-12 can then be used to determine the age of the dead matter.

Because of the radioactive decay of these trace amounts of carbon-14, we now have a totally new insight into many fascinating historical aspects over the past several tens of thousands of years. It has opened up vast panoramas previously only dreamed of by anthropologists and paleontologists, as well as others. This method was used to determine the age of the Dead Sea Scrolls at about two thousand years using measurements of the linen, which was comprised of flax. Documents found at Stonehenge, England, were dated using pieces of charcoal at the site.[6] I vividly recall the excitement some twenty years ago when the famed Shroud of Turin was believed by many to be the cloth that Christ was wrapped in while still in the tomb prior to his resurrection. However, a small piece of the cloth, which was made from flax, was measured by the carbon-14 technique, and the diagnosis proved that the shroud was made in the fourteenth century, not from the time of Christ. It was this same technique that was used to prove the authenticity of the biblical account of King Hezekiah's water tunnel—the true story[7] depicted in chapter 1.

Since the half-life of carbon-14 is 5,715 years, there is a practical limit of time for which the standard carbon dating method is accurate. This is generally in the range of fifty years to fifty thousand years, with the accuracy being plus or minus fifty years. But this can be extended over a much longer time into past history, thanks to modern instruments. It is now possible to directly detect individual carbon-14 atoms in a sample (rather than relying upon a ratio of C-14 to C-12 atoms). Molecular ions formed from C-14 atoms can be accelerated in electric and magnetic fields and then slowed by passing them through thin layers of material.[8] This allows scientists to measure as few as three atoms of C-14 out of 10^{16} atoms of C-12. Note that 10^{16} is ten million billion! Indeed, it may be this method that will be used to refine the dating of our Kennewick Man, if the courts allow further study—since only a minute amount of material is needed from the skeleton to perform such measurements.

Other radioactive techniques have been developed to measure artifacts up to several million years of age, based on the bombardment of cosmic rays that create radioisotopes of beryllium and aluminum. When an earth movement (such as a volcano, earthquake, or mud slide) buries sediments containing these metals, the cosmic ray bombardment stops. However, the radioactive isotopes continue to decay. Hence, the "clock" is set at this time. Beryllium-10 has a half-life of 1.5 million years, and aluminum-26 has a half-life of 0.71 million years. The age of the sediment specimen can be determined by measuring the change in the ratio of these two radioisotopes.

One of the most exciting recent revelations using this dating method was the discovery of the skull and left humerus bone of an extinct prehuman *Australopithecus*. This specimen was discovered in 1997 in the Sterkfontein caves of South Africa by T. Partridge and colleagues.[9] They collected specimens in the cave sediment and used the beryllium/aluminum dating technique to place the age of this specimen to be four million years.* This makes this specimen older than even the famed Lucy human fossil discovered many years ago in Ethiopia.

If we are interested in determining the age of minerals in the ground, or meteorites that have struck Earth, we can use the uranium decay chain method. Trace amounts of uranium are contained in almost all minerals in the crust of Earth. Uranium-238 decays with a half-life of 4.46 billion years, with the final stable isotope at the end of its decay chain being lead-206. Hence, the number of lead atoms found in a sample (assuming none were present originally) is equal to the loss of uranium atoms in the same sample. By measuring the number of lead-206 and uranium-238 atoms currently in the sample, the sum is simply the number of uranium-238 atoms that existed when the sample was originally formed. Knowing the half-life of U-238, we can then determine how long the sample has been in existence. The age of Earth, measured basically by this method but accurately accounting for several isotopes of lead, was determined to be 4.55 billion years.[10]

Yet another approach to determine ages of specimens ranging from fifty thousand to a few million years is the argon method. Potassium-39 is a stable isotope that naturally occurs along with a small amount of potassium-40. The latter isotope is radioactive with a half-life of 1.27 billion years (comparable to U-238). Potassium-40 decays to the stable isotope argon-40. Potassium is known to crystallize in materials of volcanic origin, thus locking in this natural ratio of potassium-40 to potassium-39. Although it is fairly difficult to determine a ratio

---

* The original discovery of ancient prehumans at this site was in 1936. Over five hundred specimens have been discovered at that location since that time. NOTE: appendix D, which contains a listing of radioisotopes commonly used in the arts and sciences (along with other uses) was generated using nuclear properties from the Chart of the Nuclides, Knolls Atomic Power Laboratory, 2002. That chart lists the half-lives of Be-10 and Al-26 as 1.5 million years and 0.71 million years, respectively. However, reference 9 lists 1.93 million years for Be-10 and 1.02 million years for Al-26.

of potassium-40 to the stable potassium-39 isotope at any point in time, it is possible to irradiate a crystalline sample with neutrons. This causes the stable potassium-39 to be transmuted into argon-39, which is radioactive (a beta emitter with a half-life of 269 years). By measuring argon-39 via the detection of the beta particles that it emits, investigators can directly determine how much potassium-39 is present.[11] Potassium-40 can be measured because it is also radioactive. Hence, the required isotope ratio can be defined, which then reveals the age of the crystalline structure.

The above technique was used to study the possible collision of an asteroid with Earth some sixty-five million years ago. This is of particular interest because there is widespread speculation that a large meteorite (perhaps four miles in diameter) collided with Earth at about this time. Such a spectacular collision would likely have caused an enormous amount of atmospheric dust—sufficient to reduce the sunlight reaching the surface of Earth long enough to suppress the plant growth needed for survival of the dinosaurs. Most scientists believe that they vanished quite suddenly, at about this same time in history, and this provides a plausible explanation.

Large impact craters and buried structures discovered in Yucatan and the state of Iowa have been studied using the potassium-argon dating method to validate the sixty-five-million-year account for such an event.[12] Trace amounts of iridium were also found in these impacted meteorites, diagnosed by neutron activation techniques. This additional piece of radiation technology provides further credence to the overall theory of how the dinosaurs became extinct.

Precious Gems

Precious gems are universally admired and desired. But few people know that radiation plays a very significant role in transforming many of them into even more beautiful and desirable acquisitions.

Basically, gems derive their beautiful hues from mineralogical impurities. Jade, for example, derives its deep green because it contains chromium. This element absorbs red, blue, and violet light waves—but not green. Beryl, which is colorless in its pure mineral form, becomes emerald with chromium impurities. On the other hand, manganese impurities in beryl turn it into a pink color.

Gemstones can be enhanced from their natural condition by several means. The principal modes of enhancement are heating, oiling, diffusion, and irradiation. Heating is a widely accepted enhancement process used on rubies, sapphires, topazes, tourmalines, and zircons. Techniques range from simply throwing gems into a fire to employing sophisticated electric or gas furnaces at specific pressures and atmospheric conditions. Oiling is an ancient process that allows oil to seep into the fissures that reach the surface of the stone. This is not a permanent process, but it is often repeated—especially during cleaning or mounting. Diffusion is a method wherein a material such as an oxide form of titanium is coated onto a stone and left there until sufficient diffusion takes place near the surface to change the color. Colorless sapphire is transformed into blue sapphire by this method.

Irradiation is also now in common use to enrich the attractiveness of many precious gems. The following is representative of how radiation is used to enhance the value of several gems:[13]

diamonds ________ changes the color from an off-white to a fancy color (e.g., green or yellow)

kunzite __________ darkens the color

pearls____________ produces blue and shades of gray ("black" pearls)

topaz ____________ changes from colorless or nearly colorless to blue; inten sifies yellow and orange shades and creates green

tourmaline________ intensifies pink, red, and purple shades

yellow beryl ______ creates yellow color

In some cases, the irradiation effects will fade away. This is the case when trying to achieve blue, orange, or yellow colors by irradiating sapphire or by irradiating beryl to attain a deep blue color. However, for the cases listed above, the noted properties become permanent.

Perhaps the most successful application of radiation to the gem industry is in the creation of blue topaz. Tens of millions of carats of topaz are irradiated annually. Electron beam irradiation is used for converting white topaz to "Sky Blue" and converting "London Blue" to "Swiss Blue" type colors.

Cobalt-60 is often the radioisotope used to create "Cobalt Blue" topaz. The gamma rays cause electrons to be released from their normal locations in the gem. The resulting color then depends upon where the electrons relocate, as well as the charge of the atoms in their vicinity. These factors control the way the stone absorbs light—thus dictating the resulting color.

Given such uncertainties, there are no guarantees of success for gemstone enhancers. As we noted earlier, trace minerals contained in the stones naturally affect gemstone colors, so the combination of trace minerals and irradiation can produce some unexpected results. Hence, many gemstones must be irradiated in order to obtain a few stones that are altered to the desired color.

Topaz is normally colorless in its pure mineral form. It is only the rare case that blue topaz can be found in nature. When cobalt-60 irradiation was first attempted to convert the much-more-abundant colorless topaz to a rich blue color, the stone changed to a cinnamon brown—and then often faded when exposed to sunlight. However, the irradiation process has since been improved significantly, and the only way to distinguish blue topaz produced by irradiation from the rare natural blue topaz is to go through an elaborate analytical process to measure the light emitted by the gemstone when it is heated. There is no way to distinguish this with the naked eye. Hence, almost all blue topaz stones now on the commercial market are produced by irradiation.

Irradiation is also used to change quartz, which is clear and colorless in its pure mineral form of silicon dioxide, into the more attractive and valuable smoky quartz. However, this transformation can be attained only when the colorless quartz contains traces of aluminum. Also, ultimate success depends upon where the electrons settle after irradiation.

The irradiation of diamonds is particularly interesting. New colors are attained in diamonds as a result of the changed crystalline structure. Given the universal attention accorded to diamonds, it is not surprising that diamond was one of the first gems to be tested with radiation. Indeed, the early tests were performed by Sir William Crookes in 1904 (the year after Marie Curie won her first Nobel Prize), when he used radium to produce gamma rays. He was able to achieve a permanent green color. However, substantially more success has been achieved in recent times with atomic particles accelerated in a cyclotron. Highly energetic protons, neutrons, or even alpha particles have been used to more efficiently obtain permanent color changes.

The stones that are irradiated today are often then heated to about eight hundred degrees Centigrade to free them of any residual traces of radiation. Indeed, the irradiation of stones can, under some circumstances, leave the gems slightly radioactive. This has resulted in a potential health hazard in some past situations. Consequently, laws have been passed that require precious-gem enhancers to allow their products to decay to a very safe level of radioactivity before they can be marketed. Some instances have occurred in the past where less-than-scrupulous dealers have placed the enhanced stones on the market before allowing them to decay sufficiently, and there have been a few cases where this has resulted in skin burns. Hence, it is prudent to purchase gemstones from reputable dealers. There should be no concern if the stone is purchased in highly developed countries, but the buyer should beware of such purchases in countries where radioactive testing is not required.

RADIATION AND THE ARTS

Radiation is even used in fields such as archeology and art. Artifacts made from materials like wood or leather will often remain intact for long periods of time if left in their natural surroundings (such as sea or mud burial). However, they will disintegrate fairly rapidly when exposed to air. Hence, radiation techniques are often used to provide long-term preservation. The ARC-Nucleart center in Grenoble, France, is an example of a conservation center where gamma irradiation is often used in two stages to preserve precious artifacts. Step one is to irradiate the artifact to kill all microorganisms that could cause decay, and step two is to impregnate the artifact with a polymer and then irradiate it to permanently harden the object. A large number of precious artifacts have been preserved in this manner, many of which can be seen in the museums of France.

Restoration work is part of the ongoing process in the world of sculpture. Natural breakage can occur—sometimes to irreplaceable masterpieces. Repairs often require the insertion of metal rods or other devices to provide required structural support. Hence, irradiating the sculpture prior to restoration procedures is sometimes done to reveal any earlier reconstruction, thereby allowing additional restoration to be done in a complementary fashion. An example of this process was the gammagraphy of a marble statue of Aphrodite, which revealed earlier consolidation work. With restoration completed, this beautiful statue now graces the Department of Greek, Etruscan, and Roman Antiquities in the Louvre.[14]

I first learned of this irradiation preservation process while on the nuclear engineering faculty at Texas A&M. Prof. Dan Reece, director of the Nuclear Science Center at the university, was fond of telling our freshman class that he participated with archeology experts at the university in helping to preserve precious artifacts from the French sailing ship *La Belle*. This "pride and joy" of the famous French explorer Robert Cavelier de la Salle sank in a storm in the shallow waters of Matagorda Bay off the coast of Texas in 1686. It was found in 1995 in eleven feet of water, amazingly well preserved in a submerged bay of mud.[15] However, once raised and exposed to the atmosphere, much of the wooden structure quickly started deteriorating. Consequently, several pieces of the structure, along with other interesting artifacts, were brought to the Nuclear Science Center for irradiation—both to kill any active bacteria and parasites and then to cross-link and harden the special polymers that were impregnated into the artifacts to permanently preserve them. Professor Reece was particularly proud of preserving the tobacco pouch that was found still intact in the shirt pocket of the first mate. How could la Salle have ever dreamed that secrets of his mission would be unraveled and preserved for posterity by a technology that would be discovered by scientists of his own country some two centuries later?

In yet another contribution to the arts, radiation is sometimes also used to authenticate rare paintings. The technique is to employ an x-ray fluorescence technique to determine the chemical constituency of the paint. By knowing the difference in paint ingredients used during differing historical periods, a definitive dating of the painting can be determined. Since paintings have occasionally been "touched up" for less-than-esthetic reasons, this same technique can quickly reveal a forgery.

Radiation technology has been used to a profound extent in helping us better understand our heritage. It is allowing us to probe millions of years into our past and then to preserve the special art and artifacts that our ancestors have provided for our enjoyment. Marie Curie must still be smiling.

12. Environmental Protection

Some readers may think a chapter dealing with ways that radiation can protect or improve our environment must be a contradiction in terms. After all, isn't radiation one of the major reasons for environmental degradation? Wasn't it the "environmental movement" that nearly shut down the nuclear industry? Didn't the Soviet Union pollute major land masses, rivers, and oceans with radioactive contamination beyond human habitation?

Indeed, cases of high levels of radioactive contamination do exist. Both accidents and the intentional disposal of radiation wastes have occurred—especially in the former Soviet Union. And yes, there have been highly publicized episodes of an "environmental" movement that have been quite negative toward the development of commercial nuclear power.

On the other hand, to many of us who have given this topic considerable thought, it is bewildering why more environmentalists are not strong supporters of radiation technology. For despite past incidents where radioactive materials were not treated with proper respect, the fact is that these are isolated events and there are methods to clean up or isolate such past misjudgments and accidents. But far more important, radioisotopes are being used in a plethora of ways to improve our environment and our ability to sustain a high quality of life—providing fresh water, protecting our oceans, combating soil erosion, and cleaning the air we breathe. Furthermore, it may be that radiation technology is the only viable environmentally compatible way to generate the increasing amounts of energy needed by a growing population over the next century, without releasing unprecedented amounts of global warming gases and heavy metals into the environment.

I'll admit that at one time (some ten to twenty years ago) I was guarded in my response when people next to me on an airplane asked about my profession. If I told them I was a nuclear engineer, about half of them would either politely look the other way or, in some cases, immediately try to arrange for a seat on the opposite wing tip!

But in more recent times, I have been increasingly impressed that by far the majority of those I meet while traveling recognize the value of nuclear technology. Perhaps it helps when I respond to the query regarding professions by simply stating that I am an environmental engineer. I can say that with a straight face, because I deeply believe that I am. I just happen to have a specialty within the environmental world; namely, a technology that legitimately holds out the promise of leaving our planet in an environmental state worthy of our grandchildren.

The American Nuclear Society has a tradition of publishing a picture of its incoming president on the July issue of *Nuclear News*. Most newly elected officers choose to be professionally photographed in a studio or at their place of work. When it came my turn, I decided to have a close colleague capture me along my favorite forest stream that flows past a campground that our family and a group of close friends have been developing for many years as a mountain retreat center. I love the outdoors—especially those spots where it is still possible to drink directly from the stream. To lose such pristine places to environmental degradation would be a tragic insult to humanity.

MAJOR ENVIRONMENTAL CONCERNS

I recognize the above opening remarks may seem a bit superficial. I feel exceptionally fortunate to be living in a country where pristine places still exist—unfettered by human misuse. But I am well aware that this is a rare privilege. If we stand back to view our global situation on environmental concerns from a broader perspective, it is instructive to take note of some very sobering facts recently reported by the International Atomic Energy Agency (IAEA):[1]

Every *second* of every day,
approximately 1,000 tons of topsoil are lost;
approximately ¾ of an acre of forestland is lost;
approximately ½ acre of arable land becomes a desert; and
approximately 1,000 tons of unwanted gases are released into the atmosphere.

Such environmental problems are compounded by population growth. The current global population of six billion is expected to climb to nine billion or possibly even double by 2100, with most of the growth occurring in poorer nations. It is further expected that up to half of this population will be concentrated in megacities of between fifteen million and twenty-five million inhabitants, thereby making it harder to find fresh water, fertile land, and clean air. Providing the energy and food needed for this burgeoning population, and doing so in a fashion compatible with environmental concerns, will be a major challenge.

Managing Freshwater Resources

Today, one in five people on Earth lacks access to safe and affordable drinking water.[2] This equates to some 1.2 billion people. Global demand for fresh, potable water is doubling every twenty-one years, according to the Food and Agriculture Organization of the IAEA.[3] Across the world today, renewable water resources available per person are roughly half of what they were in 1960. Even more sobering, this figure is expected to drop by half again by the year 2025, according to estimates of the World Bank.[4] In a closely related area of basic human need, the United Nations estimates that about 2.5 million people are currently without access to proper sanitation.[5]

Kofi Annan, UN secretary general, said in a message[6] for World Water Day in 2002 that "fierce national competition over water resources has prompted fears that water issues contain the seeds of violent conflict." He went on to say that an overarching challenge for the UN in the twenty-first century is to raise the productivity of water, to bring about a "blue revolution." Water is perhaps our most important ingredient for a decent standard of living. It is an essential commodity for life (drinking, cooking), for decent health (personal hygiene, water-related diseases), and productivity (agriculture, industry, fisheries, energy). Less than 3 percent of water on Earth is fresh, and most of it is in polar ice or too deep underground to reach. Whereas we normally think of rivers, lakes, and reservoirs as our ready supplies of freshwater, these sources comprise less than one-quarter of 1 percent of the total global water supply (i.e., one part in four hundred). By far the bulk of water on Earth is either salty or otherwise unusable, without the application of expensive treatment means. In the twentieth century, demand for potable water increased sixfold, more than double the rate of growth of the human population.[7]

Preserving Present Water Resources

Given the pressing demands for freshwater, our first duty is to preserve what we already have. Since agriculture consumes about 70 percent of the world's available water supply,[8] our first logical priority is to leverage the efficiency of water use in growing produce. We noted earlier that neutron probes are now being used to determine how much water is truly essential near the root systems of plants. This knowledge alone could substantially reduce the water demands for a given agricultural output. Radioactive tracers are also being used to determine how much water and fertilizer the plants actually require. This knowledge further reduces water demands, both for direct water use in the field as well as for use in the industrial infrastructure associated with the production and delivery of fertilizer.

Another challenge is to discern the volume flow rates of streams and rivers—since these bodies of water are often either the direct source of potable water or the source to recharge underground water aquifers.* Radioactive tracers are widely used for such measurements. Tritium (the heaviest of the hydrogen isotopes) is the ideal tracer for measuring water flows because "tritiated" water is chemically identical to ordinary water. The tritium isotope is simply substituted for one of the hydrogen atoms in H_2O. The product is called HTO. Hence, a small amount of HTO is injected into the river of interest and a sample of the water downstream (sufficiently far from the injection site to assure complete mixing) is drawn and taken to the lab for counting.[9] The flow of the river is established by multiplying the downstream sampling rate by a ratio of the count rate of the original injection over the count rate measured at the sampling point. The only disadvantage of this technique is that the beta rays emitted by the tritium are so weak that the count rate of the downstream sample cannot be measured in the field. The sample volume must be taken to a laboratory that is equipped with very sensitive counting equipment.

To circumvent this inconvenience, a total-count method is often used, in which the radioisotope injected is a gamma emitter such as bromine-82 or technetium-99m. The river flow is determined by essentially the same equation as above, but the accuracy of the measurement depends heavily on the calibration of the detector. Scientists must know how the counts per second registered by the downstream gamma detector convert to counts per volume of water. Fortunately, there are now acceptable ways to accomplish this. In practice, both techniques (i.e., tritium injection and gamma-emitter injection) are often used in conjunction to obtain maximum accuracy.

In addition to knowing how surface bodies of freshwater behave, it is vitally important to understand the performance of underground aquifers. Many regions of the world depend almost exclusively on pumping water from such aquifers for agricultural, industrial, and domestic use. When the withdrawal rate exceeds the recharge rate, considerable lowering of the water table and loss of groundwater resources may result. This not only increases the pumping power required to

* An aquifer is a subsurface zone that yields economically important amounts of water to wells.

retrieve the water; such overexploitation may induce seawater encroachment in coastal areas or allow contamination to ingress from polluted surface water bodies. Hence, it is important to understand the performance of underground aquifers to protect these resources for this and future generations. Radioisotopes are very useful in this regard.

The radioisotopes utilized for such studies are either natural or man-made. The naturally occurring radioisotopes, also known as environmental radionuclides, are classified[10] into three subgroups according to their source:

1. *Cosmogenic radioisotopes*: These radioisotopes, notably tritium, beryllium-10, carbon-14, and chlorine-36, are naturally generated in the atmosphere via cosmic radiation. Cosmic rays impact elements that always exist in our atmosphere to cause the generation of the above radioisotopes via nuclear transmutation. These radioisotopes are subsequently transported to the surface of the land and oceans through rainfall or in association with other particles.

2. *Fallout products:* These radioisotopes were produced from past nuclear bomb testing in the atmosphere. They include a range of fission products, most notably cesium-137, but also include several of the cosmogenic isotopes, such as tritium, carbon-14, and chlorine-36. Their yields reached a maximum in 1962 just prior to the 1963 Atmospheric Test Ban Treaty, but their concentrations are now substantially less because of natural radioactive decay.

3. *Primordial radioisotopes*: These radioisotopes include uranium-238 and thorium-232, along with their daughter products (such as radon). Because of the very long half-lives of these elements (the half-life for uranium-238 is 4.47 billion years and the half-life for thorium-232 is 14 billion years), these radioisotopes have been present since the formation of Earth. In addition to these radioisotopes, there is a group of stable isotopes (i.e., they do not decay with time) that includes deuterium (hydrogen with one neutron), carbon-13, and oxygen-18. Since these latter isotopes do not decay, it is not possible to use decay times for diagnostic purposes; rather, the technique is to use isotopic ratios.

The particular aspects of performing groundwater evaluations include[11]

1. identifying the sources of recharge water (those sources of surface water that seep underground to replenish the underground aquifer);

2. estimating the extent of mixing of underground water from different local and regional sources;
3. calculating the age of groundwater samples (the time that has elapsed since the surface water percolated underground);
4. determining the direction and rate of groundwater flow; and
5. understanding the processes leading to degradation in groundwater quality.

Figure 73 is included to illustrate some of the rudimentary steps that might be involved in evaluating an underground aquifer. Since surface water percolates through the soil and other overlying strata to replenish the underground water, it is essential to ensure that the recharge areas are clearly identified and protected. Environmental tritium can help immensely in such studies. For the example shown, significant levels of tritium were measured. This indicates that the groundwater has infiltrated through the unsaturated zone in "postnuclear" times—since the 1960s. We know this because most of the tritium in the environment was derived from atmospheric nuclear testing, so the relative abundance of tritium serves as a marker in time. I find it somewhat interesting that atmospheric nuclear testing (almost certainly a "blight" from an environmental viewpoint) actually has contributed to the study of something as important as our underground aquifers. This particular diagnostic capability will only last a few more decades, however, since the half-life of tritium is but 12.5 years. Hence, the abundance of this man-made tritium will eventually diminish below natural, cosmogenic tritium. Fortunately, modern detection techniques are sufficiently sensitive that this technique will likely be usable for measuring the age of recharge water for another hundred years.

Another aspect of considerable interest in characterizing an underground aquifer is to determine the age of the groundwater at different points in the aquifer. This is of interest in order to determine the fluid dynamics of the site. Such determinations can have significance regarding the volume of water available in the aquifer and thus the ability of it to sustain present and future demands.

As noted above, tritium can be used for dating the age of the water in a aquifer that is relatively dynamic. However, its short half-life limits its utility as a measuring tool to the order of a hundred years. Carbon-14 (half-life = 5,715 years) is widely used for dating groundwater samples with residence times ranging from a few hundred up to thirty thousand years. Since carbon must be dissolved in water, rather than being an actual component of water, it is necessary to understand how the dissolved carbonate entered into the aquifer. Chlorine-36 (half-life = 301,000 years) may be used to date water with resident times in excess of one million years.

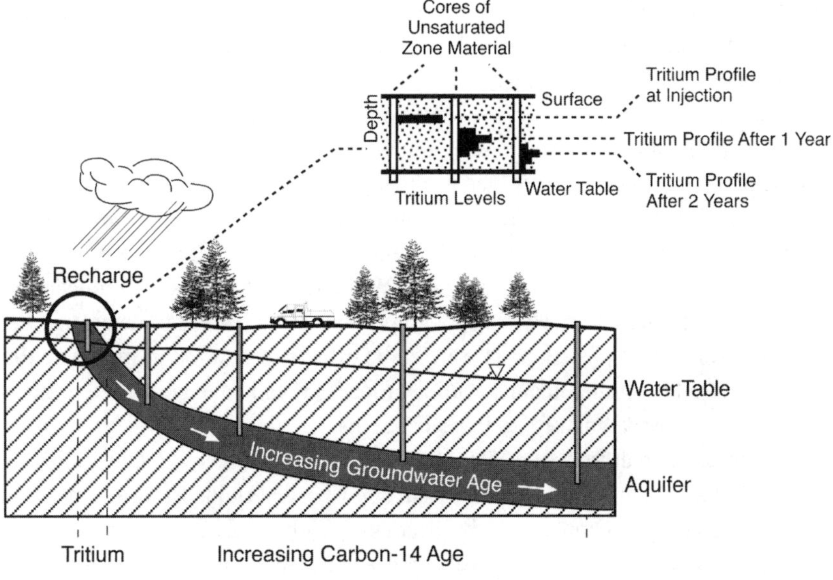

Figure 73. Characterizing an underground aquifer.[12]

The Great Artesian Basin in Australia, for example, covers about one-third of the Australian continent.[13] Understanding the behavior of this massive aquifer is of significant interest because Australia has very few rivers to serve as freshwater supplies. The major recharge area is in the east, and the water flows in a generally westerly and southwesterly direction. Measurements indicate that this basin size is large enough that the groundwater ages reach up to two million years near some of the discharge areas. Both radiological means and hydraulic modeling have been used to satisfactorily validate this conclusion.

We noted in the previous chapter that it is sometimes necessary to actually count individual atoms in making dating evaluations, rather than rely on isotopic ratios. Hence, as in the case of dating archeological artifacts, accelerator mass spectrometry (AMS) is sometimes used for environmental analyses. This is a very sensitive means of using ion acceleration along with magnetic "steering" to separate individual ions (nuclei with some of their electrons missing) so that they can be individually counted.

Preventing Contamination

Contamination of surface water is a major concern for two reasons. First, contamination seriously reduces the ability of rivers, streams, and lakes to provide usable water to support fish life, irrigation, and industrial and domestic purposes.

Second, surface water may be the primary source of recharge water for underground aquifers. Once underground aquifers are polluted, it is very difficult to purify them.

The best way to avoid surface water pollution, of course, would be to prevent any pollutants from ever reaching these bodies. However, that is often impractical—particularly in regions where high-tech effluent treatment plants are not yet affordable. Furthermore, there are many instances where the water is determined to be already polluted, but both the types of pollutants and the sources of these pollutants are unknown. Without this knowledge, remediation measures cannot be taken. Hence, it is important to have ways to determine the types and sources of pollution, as well as the effects of dilution to minimize the damage caused by contaminants. Radioisotopes can and have been used to great advantage to provide answers to these concerns.

If chemical analyses fail to determine the identity of the pollutants, identification of the elemental components can often be determined through neutron activation analysis. Samples can be irradiated with neutrons to transmute the unknown materials into radioactive species that can be readily measured and identified. A simple calculation then can reveal the original contaminant.

Determining the source of the pollutants can be done by inserting small amounts of radioisotopes at suspected locations and then sampling to see if these radioisotopes (all of which have a unique signature) are detected in the polluted region.

Understanding what types of heavy metals become contaminants is especially important. The total toxicity of all metals mobilized by human activities each year exceeds the total of all radioactive and organic wastes combined.[14] These metals include arsenic, cadmium, copper, lead, and mercury—all of which are easy to detect using nuclear-based analytical techniques.

One study[15] of water polluted with toxic metals was carried out by the IAEA in Brazil, where poor people used mercury to extract small quantities of gold left in mine tailings along the Amazon River basin. There the people made no effort to reclaim the mercury; rather, they either evaporated it into the air or dumped it into rivers. The toxicity of mercury is of concern even at parts per billion, so anyone living in the vicinity who consumes fish or uses the polluted water for domestic or agricultural needs is at risk. Hence, the IAEA measured the mercury content in human hair, fish, and river sediments in the affected areas using sensitive neutron activation analyses and provided its results to local authorities for proper action to this complex problem.

The effects of dilution are significant to determine, because discharges of contaminants into surface bodies of water may be entirely acceptable if done under carefully controlled conditions. A relatively straightforward procedure is to inject a gamma emitter such as gold-198 (half-life = 2.7 days) into the body of water such as a river and then measure the plume downstream with a detector pulled by a survey vessel. By traversing an area in a lattice fashion, and also raising and lowering another detector in a uniform manner, the survey vessel can gather sufficient information to provide a three-dimensional pattern of the polluting plume versus time.

The effects of discharging treated sewage into a body of water can be determined by recognizing that this sewage is comprised of a mixture of dissolved contaminants, suspended particles, and dispersed grease. By using appropriate radionuclides, these three components can be independently tagged and studied separately. For example, the dissolved components can be labeled with tritiated water (HTO), the fine particles with gold-198, and the organic grease with tritiated organic compounds.

Obtaining New Sources of Freshwater

Certainly there is incentive to find new sources of freshwater, in addition to protecting and preserving the existing sources. But with most of the readily available freshwater sources already tapped, and both the population and per capita water use on a rapid climb, responsible planners are now searching for fundamentally new ways to obtain potable water.

Given that over 97 percent of the water on the planet is seawater, and that many major population centers are situated along the coastline, an obvious target is to explore desalination (i.e., extract the salt and produce drinkable water). The concept is not new. Indeed, there have been several small desalination plants in operation around the world. The simplest method is to heat the saltwater sufficiently to vaporize the water, leaving the salt behind. Only a source of heat is needed. The catch, as always, is the cost. The energy required to produce potable water by conventional sources is in most cases prohibitively high. Furthermore, if the heat is supplied by burning fossil fuels, the environmental problems, owing to expelling pollutants such as CO_2 plus sulfur and nitrogen oxides into the atmosphere, could be unacceptable.

Nuclear reactors have been used for such purposes in the past. The BN-350 reactor, located in the former Soviet Union (now Kazakhstan), was operated for many years as a dual-purpose reactor—generating 150 MW of electricity and over one million gallons of desalted water per hour. New nuclear reactors are now being designed that can be as economical to operate as the most efficient fossil-fueled desalination plants, according to the president of the Association of Atomic Engineers of Morocco.[16] One design is reported to be capable of producing two cubic meters (over five hundred gallons) of water for a dollar. If so, this is getting close to economic viability and certainly worth the cost in many regions of the world today.

India has built a demonstration desalination plant in southeastern Kalpakkam, which was expected to begin operating in 2003.[17] This facility has a capacity of 1.7 million gallons per day and is linked to a nuclear reactor. Whereas this plant will not have the cost effectiveness of a new design, it should provide valuable experience for members of the project. More than twenty nations are now exploring options for nuclear-powered desalination operations through IAEA programs.

GUARDING THE OCEANS

Approximately 70 percent of Earth's surface is covered with saline seas.[18] More than half of the world's population lives along coastal zones. Nearly one billion people, mostly living in developing countries, depend upon fish as their sole source of protein. Hence, the oceans of the world are becoming increasingly recognized as precious commodities—worthy of special attention for environmental protection.

Environmental isotope techniques are very well suited to study the dynamics of the large and complex systems associated with oceans. There are basically two classes of ocean investigations that are of particular interest.[19] The first concerns processes within the coastal zone, where impacts arise from rivers that discharge contaminants resulting from agricultural, industrial, or urban-development sources. The second concerns circulation patterns within the deep oceans and their impact on weather, climate change, and the support of marine life.

There are two groups of radionuclides that are naturally present in the oceans (all at very low concentrations). The first group includes tritium and carbon-14. They enter the ocean via rainfall and by exchange with the carbon-14 in the atmosphere. The second group includes uranium and thorium, along with their radioactive daughter products. These constituents enter the coastal zone either from being dissolved in river water or from being embedded in discharged river sediments.

One of the largest, coordinated international efforts to date—in assembling the staff and equipment dedicated to the assessment of oceanic environmental issues—is the IAEA's Marine Environmental Laboratory (MEL) located in Monaco.[20] Numerous studies have been initiated from this laboratory, wherein the latest radiation tools are put to use to decipher sources of pollution.

Coastal Zone Concerns

Activities on land constitute the source of some 80 percent of all marine pollution. These include sewage and industrial effluents, fertilizer runoff, heavy metals discharge, and persistent organic pollutants. Radioisotopes provide unique methods to determine the sources and types of pollutants that cause the greatest damage.

One example of a key body of water becoming heavily polluted is the Black Sea.[21] It was declared nearly dead a decade ago, and dubbed by the *Washington Post* as "the toilet bowl for half of Europe." More than sixty plant and animal species essential to the Black Sea ecosystem, including dolphins and seals, are endangered or nearly extinct, as well as thirteen types of commercial fish encompassing many species. Waterborne diseases are common all along the coast, and outbreaks of cholera have caused beaches to be closed in numerous locations.

This body of water serves as the major natural and economic resource for over 160 million people. The Black Sea basin includes major parts of seventeen nations and thirteen capital cities, including the six immediately surrounding countries of Bulgaria, Georgia, Romania, the Russian Federation, Turkey, and Ukraine. The second, third, and fourth major European rivers, the Danube, Dnieper, and Don, discharge into this sea.

Given the strategic importance of the Black Sea, there has been considerable international collaboration within the past decade to study the reasons for the pollution and ways to restore it to its former health. The IAEA has supported over fifty research cruises throughout the Black Sea, including two multinational scientific cruises, to gather information about the environmental status of the sea. Radioactive techniques have been the major tools used in these activities to diagnose the underlying causes for the pollution. Radiological tracing techniques have been employed to both assess the fate of the pollutants as well as to understand the marine processes taking place.

Given the factual basis that now exists for understanding the demise of the Black Sea, leaders of the participating scientific groups believe the sea can be saved. This diagnostic effort, largely successful because of the utilization of radioisotopes, is expected to yield an immensely positive impact on the future of the region.

Another example is the incidence of "red tides" in the Philippines, where toxic algae blooms have caused massive paralytic shellfish poisoning.[22] Given that approximately seven thousand islands spread across thousands of square miles of tropical seas constituting the Philippine Islands, it should be a fisherman's wonderland. Indeed, this has been the case for many decades. However, in February 2002 hundreds of tons of dead milkfish suddenly started floating to the surface in Bolinao. Whereas this episode was perhaps the most dramatic, such incidents have been occurring over the last several years in some seventeen coastal areas of the Philippines, causing devastation to the local fishing industry—and posing a health hazard to the citizenry.

The standard, if somewhat primitive, technique to determine if the shellfish toxin concentrate exists in a coastal area is to inject lab mice with samples of the suspected concentrate and see how long it takes for the mice to die. Needless to say, this method is highly inaccurate and time consuming—often causing unnecessary harvesting bans even though the fish are perfectly safe for human consumption. Better methods of diagnosis are clearly warranted.

A new technique being tested by the IAEA for early diagnosis is a receptor-binding assay technology. This approach relies on the radioactive properties of iodine-125 as a labeling mechanism to detect abnormalities in growth hormones of the suspected shellfish. Considerable excitement now exists among leading scientists at the University of the Philippines that this new nuclear technique will become the standard for ensuring safety of the public while simultaneously supporting the local fishing industry.

Understanding Our Deep Seas

Both tritium and carbon-14 have proven to be particularly useful radioisotopes in studying mixing processes in the upper ocean. By virtue of numerous profiling studies, a fairly complete understanding of oceanic circulation patterns has been established. Tritium and carbon-14 fallout from atmospheric nuclear testing are particularly useful for measuring ocean currents down to a depth of about a half mile. For the study of currents at deeper depths, reliance upon the cosmogenic sources for these radioisotopes is necessary.

An understanding of ocean currents is very important in the determination of climate changes. For instance, there is general agreement that the Gulf Stream is responsible for the large difference in winter temperatures between Boston and Rome, even though both cities are at the same latitude. Normally, the Gulf Stream soaks up heat in the tropics and then moves up the East Coast of the United States and Canada, heating up the air and providing warmth to that region. However, because the prevailing North Atlantic winds blow eastward, much of that heat ends up in Europe. This can cause Rome to be as much as thirty-six degrees Fahrenheit warmer than Boston in the winter. Having given up its heat, the Gulf Stream water in the Northern Hemisphere then becomes colder and denser, and this huge mass of water sinks into the North Atlantic by a mile or more in a process oceanographers call thermohaline circulation. This provides the "engine" to drive the cold seawater back to the tropics for another cycle.

However, scientists at the famed Woods Hole Oceanographic Institution, located at Cape Cod in Massachusetts, are beginning to fear that massive melting of the polar ice caps caused by global warming could nullify this natural ocean cycle by flooding the North Atlantic with less dense freshwater and preventing the Gulf Stream from sinking in that region.[23] If this should happen, the Gulf Stream could slow down or even reverse direction, and the Woods Hole scientists are concerned that this could happen rather suddenly. They fear the possible return of a localized Little Ice Age, where temperatures along the East Coast of the United States and the regions north into Canada could drop significantly for several decades. If this is a real possibility, being able to measure any changes in the Gulf Stream flow could be very significant. Radioactive techniques are still probably our best way to gather this information.

There is still considerable controversy among the scientific community about the actual effects of global climate change. But the big problem is that if major climate changes should occur, there is very little (if anything) that we can do about it. The consequences, whether a substantial heating or cooling effect, could be devastating.

Understanding Soil Erosion

In addition to having ready access to freshwater, fertile soil is central to sustained development in a world where the population continues to expand. The demands for increasing food production dictate that productive farms remain a high priority. Hence, protecting good topsoil from wind and water erosion is a vital matter.

Loss of Topsoil

In recognition of this problem, substantial efforts have gone into the collection of groundcover data via satellites. However, there is a need to validate or interpret the imaging data with independent estimates of erosion patterns. One of the primary methods of conducting these studies is to measure the distribution of cesium-137. We recall that this is a fission product widely dispersed in a fairly uniform manner by atmospheric bomb testing mostly in the 1950–1960 time frame. This radioisotope has subsequently settled to the surface of Earth where it has become adsorbed in clay and other natural topsoil constituents. Since it has a half-life of about thirty years, cesium-137 supplies scientists with several decades for investigating areas where its presence is reduced and regions where it has accumulated, thereby revealing surface erosion patterns.

The rate of surface erosion depends upon a number of parameters, such as the type of soil and its distribution, the amount and intensity of rainfall, the slope of the land, the nature and extent of vegetation, and the erodability of the soil.

Many attempts have been made to study erosion with mathematical formulas, but there are still substantial uncertainties using such an approach. Thus, having the ability to make direct determinations via measuring redistribution patterns of cesium-137 provides a major environmental tool. Such detailed studies can be very localized. For instance, in one vineyard study, it was found that approximately one bottle of fertile soil was lost to the cultivated area for each bottle of wine produced.[24] Though possibly bad news for the owner of the vineyard, I suspect there are several readers who might be willing to trade a bottle of dirt for a comparable volume of fine wine!

We might note that using cesium-137 as a dating technique for soil erosion is fundamentally different from that used for the majority of radiometric methods. This technique does not depend upon radioactive decay; rather it depends upon the relative changes in total radioactive activity as detected in the eroded versus the accumulation sites. Other more-conventional radioactive methods can be used for dating erosion patterns, however. If our desire is to measure sediment accumulation over time scales up to a few decades, we can use lead-210, which has a half-life of 22.3 years. If our interest is erosion patterns up to tens of thousands of years, we might use carbon-14 (half-life = 5,715 years). Finally, if our interest is over the very long archeological time of hundreds of thousands of years, we can use the ratio of thorium-230 to uranium-234. The half-lives of these two radioisotopes are 75,380 years and 245,500 years, respectively.

Erosion along Waterways

Considerable attention is now being given to the erosion of sand and sediment along rivers and oceans. This is especially important to commercial businesses that line our waterways, as well as to homeowners who build along scenic bodies of water. It is also vital to agricultural specialties such as clamming.

To investigate the transport of sand and sediments, it is necessary to match the particle size distributions and density of the natural materials with those of the radioisotope used as the tracer.[25] The method generally used for investigating sediment erosion is first to collect samples from the area of investigation and make it up as a slurry. Then a radioactive tracer that will strongly adsorb to the surface of the sediment particles, such as gold-198, is added. The mixture is then carefully returned to the site, and radiation detection counters are used to monitor the spread of the mixture as a function of time.

The method normally used to investigate the erosion of sandy beaches is to synthesize glass beads having a size distribution and density matching that of the sand. The glass is impregnated with an appropriate target material (such as lanthanum oxide, iridium, or silver) and then irradiated in a nuclear reactor to form the desired radioisotope tracer (e.g., lanthanum-140, iridium-192, or silver-110m). The radioisotope chosen depends primarily upon the proposed length of the study. For a study lasting only a few days, lanthanum-140 would be a good choice (half-life = 1.7 days). If the study is to last a full week, gold-198 might be used (half-life = 2.7 days). If the study is to last one or two seasons, iridium-192 is probably a good candidate (half-life = 73.8 days). Finally, if the study is scheduled for a complete annual cycle, silver-110m provides a good choice (half-life = 250 days).

Erosion within Waterways

Finally, there continues to be interest in determining the accumulation of sand or sediment behind dams, shipping channels (rivers or canals), harbors, and the ocean itself. This information is needed to determine the useful lifetime of a dam, the degree of dredging that may be necessary to keep shipping channels open, and the potential danger from shallow ocean navigation.

The methodology for conducting these studies is quite similar to that of evaluating the erosion of sandy beaches, although the means of detecting the transport of the particulate matter may be a bit more complex, since essentially all of the measurements must be obtained under water. However, well-equipped survey vessels are available for such work.

Again, the particular radioisotopes chosen for the task depend primarily on the length of the proposed study.[26] Gold-198 (half-life = 2.7 days) is often used for the one- or two-week studies. Chromium-51 (half-life = 27.7 days) is popular for studies up to about four months, iridium-192 (half-life = 73.8 days) is often used for eight-month studies, scandium-46 (half-life = 83.8 days) can be used for studies up to about nine months, and silver-110m (half-life = 250 days) is generally the workhorse for studies over one year.

Radiological Contamination

We acknowledged at the opening of this chapter that radiation contamination has occurred at several sites of our globe since the dawn of the nuclear age. Substantial quantities of high-level nuclear waste have been generated since the discovery of nuclear fission in the late 1930s, and it has not all been dealt with in a proper manner. Much of the generation of this waste is the result of the nuclear weapons industry, and it is within that infrastructure where most of the contamination has taken place. However, substantial amounts of radioactive waste have also been generated in the commercial sector, and even though very little harmful contamination has taken place in this arena, the amounts of waste that have been generated certainly warrant our attention for environmental scrutiny.

Nuclear Weapons Production

Shortly after the discovery of nuclear fission, innovative scientists recognized that it might be possible to transform this enormous energy into powerful nuclear bombs. The efforts expended to actually build these weapons were extensive, and because of the military implications, they were all conducted under the utmost levels of secrecy. The US effort was conducted within the Manhattan Project, where the original plutonium production site was at Hanford, Washington (and later Savannah River, South Carolina), and the uranium enrichment activities were initiated at Oak Ridge, Tennessee (and later at Portsmouth, Ohio, and Paducah, Kentucky). A special assembly plant was set up at Rocky Flats, near Denver, Colorado, and the early naval nuclear program was established near Idaho Falls, Idaho. All of these sites generated significant amounts of nuclear waste, but the bulk of the high-level nuclear waste (generally defined as reactor-generated) was produced at Hanford and Savannah River.

The largest amount of high-level nuclear waste accumulated within the United States for the military effort occurred at Hanford. As such, some have referred to this site as the most polluted site on Earth. However, I have lived immediately adjacent to the Hanford project for over thirty years and I know of only one person who has demonstrably suffered any adverse health effects due to site activities (which now extends to approximately sixty years). That one incident involved contamination spread in a work area associated with a chemical explosion in a glove box. This is not to say that no other ill effects have occurred;* rather, it is to provide some perspective to counter many of the horror stories that continue to circulate from those either uninformed as to the medical facts of the matter, or who simply choose to spread fear because of the notoriety they can achieve by the willing media that recognize the commercial value of fear-producing headlines. The principal reason that the actual health hazards from the Hanford nuclear production history are low is that the dangerous wastes are largely contained in huge underground tanks. It is true that some of the older

---

* Numerous scientific studies have been conducted to asses the potential health effects of radiological contamination to the air and the Columbia River over the half century of military operations at Hanford. Whereas these studies have consistently concluded that there are no large-scale health impacts, it is not possible to statistically rule out any individual health problems.[27]

tanks have leaked, and it is likewise true that substantial amounts of lower-level radioactive waste were purposely dumped into open ground trenches during peak plutonium-production periods some forty years ago. However, strict environmental monitoring has been practiced since that time. For an even-handed and comprehensive perspective on the environmental aspects of the Hanford site, the reader is referred to an excellent book by Roy Gephart, entitled *Hanford: A Conversation about Nuclear Waste and Cleanup.*[28]

By far the most serious radiological insults to the environment occurred within the former Soviet Union (FSU). Don Bradley, in *Behind the Nuclear Curtain: Radioactive Waste Management in the Former Soviet Union*, has done a marvelous job in categorizing the contamination that took place during the heavy weapons production years in the FSU.[29] By comparison, the sites in the United States look pristine. In many cases, high-level nuclear waste in the FSU was dumped directly into surrounding lakes and rivers. The Techa River and Lake Karachai, both located in the Ural Mountain region, are still so radioactive that the waters will not be safe for appreciable human use for several more decades. An accident occurred at the Mayak site in 1957 when a radioactive-waste tank became so dehydrated that it exploded, casting radiation over a wide area. Much farther to the south, the Chernobyl reactor suffered a major accident in 1986 after technicians deliberately disabled all the safety systems and pushed the reactor into a mode that they had never tried before in order to conduct a special experiment. The contamination was, indeed, very widespread. However, the concentration of contamination produced by that widely publicized event did not result in nearly the levels experienced at the nuclear weapons production sites.

Fortunately, the population density at the most contaminated FSU sites is quite small. Hence, a very small fraction of the global population will have any access to the contaminated sites. This is not to excuse the unconscionable abuses taken in the FSU, but the fact is that relatively few people (beyond those workers who were forced to stay on the job during the height of the cold war) will suffer any consequences.

One other radiological contamination issue associated with the military concerns the nuclear-powered submarines that the FSU deliberately sunk and discarded in the northern oceans. Most of the focus has been on the relatively shallow waters of the Kara and Barents seas on the northern coast of Russia. Studies carried out by the IAEA Marine Environment Laboratory (MEL) in Morocco have revealed clear traces of the buried submarines and their discarded fuel loads, but the detectable radiation was low and determined to not be dangerous from a radiological point of view.[30] Water provides an excellent shield for radiation. Hence, the only serious concern is whether substantial migration of the contaminated sediment on the ocean floor might occur. Fortunately, the movement of such materials has been shown to be very small. The principal radiation detected was determined not to be from the sunken submarines, but rather from weapons fallout and discharges from the Ob and Yenisey rivers (a manifestation of the contamination from the weapons production sites mentioned above).

Commercial Nuclear Power

The amount of high-level nuclear waste generated by commercial nuclear power plants around the globe now exceeds the amount generated in the weapons programs during the cold war. However, this waste is very carefully stored and regulated. I am not aware of any significant hazard that has been inflicted upon any member of the public from such material.* As discussed in chapter 6, the ultimate disposition of high-level nuclear waste is still a hotly debated topic in many regions of the world. But the degree of regulatory oversight associated with commercial nuclear power provides strong guarantees that the actual health hazards of handling this material will be exceedingly small.

Whereas the commercial nuclear waste issue is far too large a topic to cover adequately in this book, I might note that I have personally stood on top of the shielded, encapsulated high-level waste packages at both the La Hague reprocessing plant in France and the Sellefield reprocessing plant in Britain, and my radiation dose rate was *less* than normal background. Exceptionally tight controls, including all of the relevant shielding packages, are rigorously built into the handling standards for this material. In fact, the publicly published records of these plants provide clearly recorded data showing that the workers in these plants get *less* radiation during their work shifts than they would receive if they were simply walking around outside during the same workday.

Alarmists like to scare the public by claiming that shipping the nuclear waste packages to a final repository places innocent people in harm's way. But actual data show just the opposite. As of 1994, some two million shipments of radioactive waste were being made annually in the United States, with by far the bulk of it being of very low radiological activity.[31] By way of perspective, for every shipment of these radioactive materials, there were five thousand shipments of other hazardous materials (such as flammable liquids, corrosive materials, etc.). There is not a single recorded incident of any member of the public being exposed to high levels of radiation or being injured because of a radioactive release from any of these packages.

*The Chernobyl accident contaminated a very large area. However, this contamination resulted from an accident while the reactor was running, rather than a consequence of handling the nuclear waste materials.

Very few of these shipments contained high-level waste because most of that material in the United States is still stored at the sites of the nuclear power plants. But there should be no undue alarm attached to their eventual transport. The degree of safety that goes into high-level shipping casks is extraordinary. In order to receive a license for use, these casks must undergo a harrowing series of tests, including their demonstrated ability to survive a thirty-foot free fall onto an unyielding surface, a forty-inch fall to strike a six-inch-diameter pin, a thirty-minute exposure to a searing fire of 1,474 degrees Fahrenheit, and complete immersion in water for a period of eight hours—all without a leak of any of their contents.

I am personally confident that technical solutions exist to handle all of the radiation health hazards associated with the commercial nuclear power industry in a manner such that the benefits substantially outweigh the risks. The only question is how long it will take for the industry to demonstrate solutions sufficient to satisfy a skeptical public.

Polluting Our Atmosphere

Clean air has become a rallying cry among a large segment of the public ever since smog patterns began to emerge in major cities, such as Los Angeles, Houston, Mexico City, and other large metropolitan areas around the globe. Certainly this is something to be concerned about, and it is not a new problem. One of the more notorious public disasters occurred in London in 1952, when air pollution due to coal burning caused thirty-five hundred more deaths than normally expected in a span of only a few days. Today, atmospheric pollution comes from many sources, including industrial emissions, car and truck exhausts, and coal and wood combustion. The Earth Summit in Rio de Janeiro in 1992 focused international attention on the need to curb atmospheric pollution, and it was followed up by the Kyoto Earth Summit in 1997. Major goals were set to reduce carbon emission owing to concerns over global climate change. Unfortunately, progress to meet these goals has been slow.

Major Pollution Concerns

One of the most gripping images that I have seen regarding atmospheric pollution is the "Brown Cloud" now hovering over the southeastern Asian continent, stretching from India to China. I have included this as figure 74 because of its impact. As reported by Reuters[32] on August 11, 2002, and in the *London Times*[33] on August 12, 2002, this cloud was about two miles thick and threatening the lives

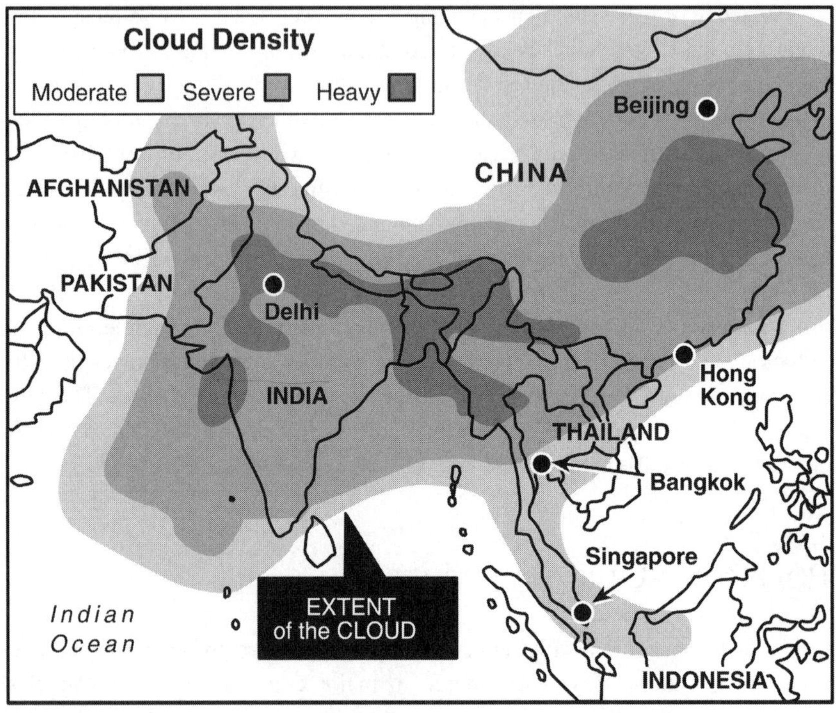

Figure 74. The Asian Brown Cloud in August 2002.

of millions of people in the affected region. Approximately 80 percent of the haze was reported to be caused by human activities (burning fossil fuels in vehicles and power plants, industrial effluents, and emissions from millions of inefficient cookers). The remaining 20 percent was reported to be due to forest fires. This haze is already altering regional monsoon and rainfall patterns, as well as causing acid rain that damages crops and trees, and threatens hundreds of thousands of people with respiratory disease. Perhaps most sobering, many believe both regional and global impacts of the haze will intensify over the next thirty years as the population of the Asian region rises to an estimated five billion people.

Of special concern to human health are small airborne particles (less than ten micrometers) that can penetrate the lungs, causing respiratory or heart disease and even death. Consequently, the IAEA has set up a global network in major cities to collect air samples and measure the concentrations of these pollutants using nuclear-based techniques. The purpose is to build a composite picture of the particulate sources to enable health and environmental authorities to devise mitigation strategies.

In a study of ten cities in the Asia-Pacific region, healthy nonsmokers have participated in a program where they were asked to inhale a special vapor tagged with technetium-99m and then allow their exhaled breath to be measured by a special camera-computer system.[34] Residents of all but two cities showed signs of injury to the lung membrane that separates air and blood. These findings demonstrate the severity of urban pollution problems, helping to explain increases in respiratory problems in children and cardiovascular disease in the elderly.

Global Climate Change

Certainly the most widely publicized aspect of atmospheric pollution within the past decade is global climate change. Many people equate the words *global climate change* with *global warming*. I prefer not to do this, because I am not an atmospheric scientist and the rhetoric going back and forth on this issue is somewhat akin to watching a Ping-Pong match. I simply don't know if we are now on an irreversible path that spells gloom and doom to our descendants. But this I do know. The correlation between average atmospheric temperature and atmospheric carbon dioxide concentration is real.

It was while I was still mulling over all of this that Prof. David Scott of the University of Victoria (Canada) provided me a graph[35] illustrating the levels of carbon dioxide concentrations over Antarctica during the past four hundred thousand years.* The temperatures over Antarctica during this same time period correlated so well with the carbon dioxide variations that the two were hardly distinguishable. The insert to figure 75 illustrates the carbon dioxide variation (with the temperature variation essentially identical). Most striking to me is that at no time during this four hundred thousand-year period did the carbon dioxide levels exceed 325 parts per million.

Now take a look at the global atmospheric carbon dioxide levels over the past 1000 years, shown on the lower portion of figure 75 graciously provided to me by Dr. Hans-Holger Rogner of the IAEA.[36] We see that the global CO_2 level remained at approximately 280 parts per million up until about the year 1800 and then increased slowly until it moved up sharply at about the year 1950. Today the number is about 380 parts per million. Stated differently, the CO_2 level in our atmosphere is now well above any level that existed in our past four hundred thousand years, and there is no end in sight. The range from the year 2000 to the end of the present century is bracketed by several modeling assumptions, in which many variables have been incorporated in an attempt to estimate the range of levels that may actually be reached. When I compare these two graphs, I become *very* concerned!

* These data could not have obtained without the application of radiological dating techniques.

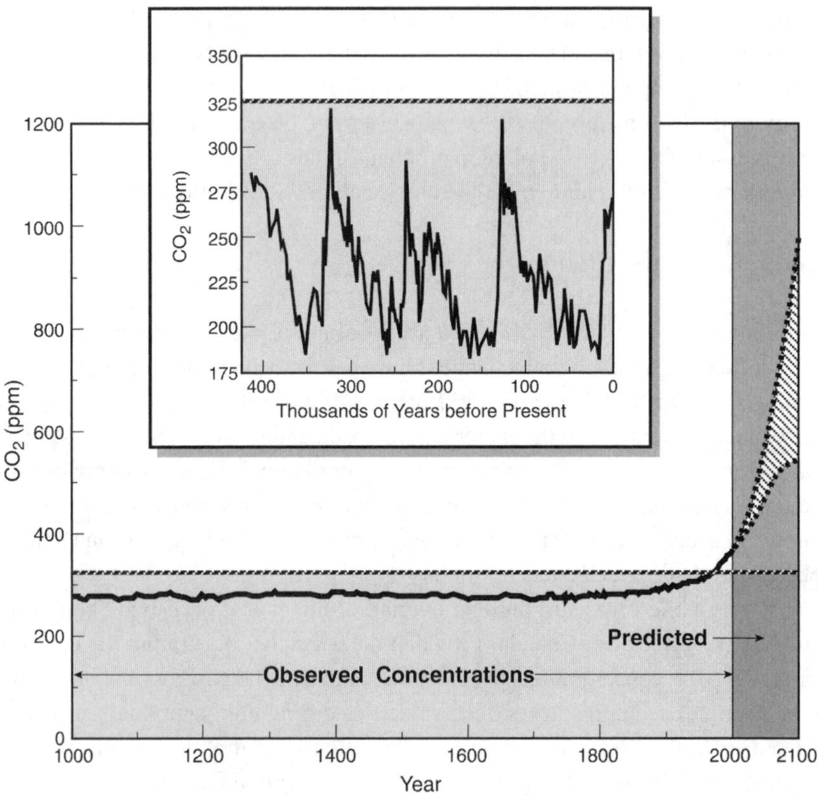

Figure 75. Evidence that carbon dioxide concentrations in the atmosphere are now higher than in the last four hundred thousand years, and increasing dramatically.

Most scientists today believe that this rapid rise in CO_2 levels will unmistakably lead to substantial warming of our planet. Yet the discussion earlier in the chapter, based on the observations of the Woods Hole Observatory, suggests (somewhat ironically) that the heating process might actually cool some areas of the globe substantially because of a massive change in ocean currents. It is because of complexities of this type that I am reluctant to stake my professional reputation on the line in terms of just what major increases in global CO_2 levels will actually cause.

But my bottom-line message is clear. Substantial climate changes *will* very likely occur, and once they begin, there is essentially nothing we can do to reverse them—short of several generations. This, to me, is a very sobering conclusion. Hence, my plea is that we take aggressive action now to slow down our massive infusion of CO_2 (and other so-called greenhouse gases) into our atmosphere. We have other choices, including solar, wind, and nuclear technology, to effect such reductions, and I believe we have no choice but to pursue them with considerable vigor.

Energy and the Environment

We noted in the chapter on electricity that we will need to find ways of generating huge amounts of energy for the foreseeable future. Whereas there are several factors that impact the quality of life, there is perhaps no single entity that has more bearing on achieving high levels of a quality life than energy. With energy comes electricity, sanitation, fresh water, and jobs. This is dramatically pointed out by comparing energy usage per capita as a function of economic prosperity (consider, for example, figure 2 of my *America the Powerless: Facing Our Nuclear Energy Dilemma*).[37] The question is how will we be able to generate the huge amounts of energy needed while at the same time avoiding the pollution patterns that we have just discussed.

Figure 76 is included to illustrate the principal greenhouse gases generated by eight energy sources. The pollution is plotted as grams of carbon-equivalent per kilowatt of electricity produced. The data are taken from Joseph U. Spadaro, Lucile Langois, and Bruce Hamilton's "Assessing the Difference: Greenhouse Gas Emissions of Electricity Generating Chains,"[38] where the plots for 1990 technology are provided as the average of the high and low referenced values. For the fossil fuels and solar energy, the estimates, from the referenced source, are also shown for the reduction in pollution anticipated by advanced technology that may be available. Also for fossil fuels, the carbon-equivalent is broken down into stack emissions and other chain steps. The latter consists of pollution caused when mining the fuel, producing the materials required to build the plant, decommissioning, and so on.

We readily note that the nonfossil sources do not emit any pollutants up the stack. Rather, all greenhouse gas pollution liabilities for these sources comes from other chain steps.

Coal-fired plants are by far the worst from an environmental perspective, followed by oil and natural gas. Natural gas plants do not produce sulfur oxides, and the CO_2 output per unit of energy produced is less than that for other combustible sources. But it is still far more of a polluter than any of the nonfossil sources.

From this illustration we note that nuclear and hydroelectric power are the least polluting sources (a factor of fifty to a hundred times less than coal). The relatively large value for solar photovoltaic (solar PV) is due to the toxicity of the materials

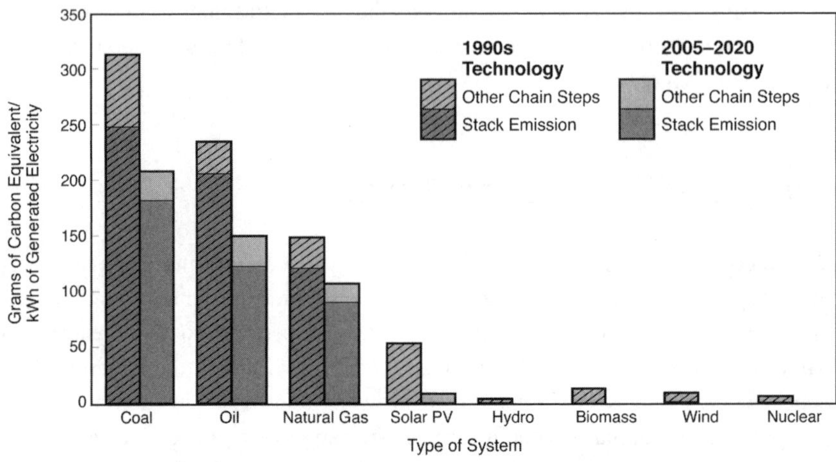

Figure 76. An environmental comparison of pollutants discharged into the atmosphere by various electricity-generating sources (all 1,000 MW equivalent electrical plants).

required for the PV system. However, improved technologies are expected to result in considerably less waste from production in the next decade or two.

Whereas nuclear, hydro, solar, wind, and biomass appear to be the winners if our concern is the quality of the environment, the other question we must ask is which of these five have the capacity to produce the huge amounts of power that will be needed. Hydro, solar, wind, and biomass would all require very large land masses to do the job. Hydro could possibly be more efficient in terms of land mass requirements for unique sites, but very few such sites remain. Although both wind and solar offer substantial promise, they are inherently unreliable, because of the day/night solar cycle and changes in weather that cause intermittent performance for both sources. Consequently, backup energy sources would be needed for these options. Biomass success requires substantial breakthroughs in cost and viability of carbon sequestration.

It is at this point that nuclear looks so attractive.[39] Both small land masses and small quantities of fuel are needed to operate nuclear plants. Hence, they should be able to provide essentially all the power necessary for an expanding global use for at least the next millennium.

In reality, we will very likely need all of the energy sources that we can viably build and operate. But we must drastically reduce our dependency upon fossil fuels, both because of the certainty of eventually running out of these finite resources and because of environmental concerns. Hence, once again, the technology resulting from Marie Curie's marvelous discoveries may become even more essential as we power our future in a manner consistent with environmental stewardship.

13. Modern Economy

Anyone who has attempted to write a successful book knows the importance of starting out with a bang. The obvious reason is that if the first chapter does not set the hook, the remaining chapters will never be read.

Following this logic, I was very tempted to begin this book with the present chapter—because of the reaction that I experienced when I first learned of the economic impact of radiation on the US economy. It caught my attention like nothing else that I can recall in my professional career. I was simply dumfounded.

To provide some perspective, I did not choose my profession because of what it would mean to the bottom line of some corporate quarterly report. Rather, I pursued degrees in nuclear engineering because of the field's scientific importance, combined with my conviction that it promised substantial value to improve the quality of our lives. I had no idea at the inception of my career that radiation technology could productively contribute to such a large cross section of our day-to-day standard of living.

Hence, when I first learned of a study performed in the early 1990s revealing that radiation technology was contributing *$330 billion* and *4.1 million jobs* to the US economy, I simply did not believe it. Engineers, as many of you know, are often dubbed with the label "Doubting Thomas," because they need hard proof before supporting any proposition that goes beyond their experience base. But I was so fascinated with the reported results that I dug into them, found them to be true, and subsequently felt compelled to share the story with others.

The data presented below are for the United States, Japan, and China. I wish

I had comparable data for all nations, but I simply have not been able to find such material (if it exists at all). Nevertheless, grasping an overview of how radiation has impacted the national economy and job market for these major nations should provide ample evidence of how important Marie Curie's discoveries have become in our everyday lives.

UNITED STATES

Roger Bezdek and his team[1] sent a lightning bolt through the scientific community when they revealed the economic impact of radiation technology within the United States for 1991. In that year, $73 billion was added to the gross domestic product (GDP) of the United States by nuclear power, and $257 was contributed by the use of radiation. The grand total of $330 billion was nothing short of startling. It represented about 5 percent of the GDP for the United States. Furthermore, the 1991 study revealed that nuclear power created 400,000 jobs, and the nonpower radiation technology created another 3.7 million jobs, for a grand total of 4.1 million jobs.

Whereas I was initially surprised enough to see these bottom-line figures, I was even more surprised to note that radioisotopes have a substantially larger impact on the US economy than nuclear energy. After all, I grew up totally engaged in the design and analysis of advanced nuclear reactor systems, and the world of radioisotopes was but a distant blur. At this point, however, I fully realize why the contributions of radioisotopes are so large. Indeed, most of this book is devoted to the plethora of these applications beyond the more obvious sector of nuclear energy.

Mr. Bezdek updated his studies a few years later[2] to include an even more detailed analysis to articulate the impacts of radiation technology in the United States for the year 1995. Table 4 contains these even more impressive results.

Table 4. Economic and job impacts in the United States from radiation technology in 1995

	Sales (billions)	Jobs (millions)	Taxes (billions)
Radiation*	$330.7	3.95	$60.9
Nuclear Energy	$ 90.2	0.44	$17.8
TOTAL	$420.9	4.39	$78.7

*Nonpower applications of radiation technology

Yes, you read it right. In 1995, radiation technology contributed $421 billion to the GDP of the United States, along with 4.4 million jobs and $79 billion of tax revenues. To the best of my knowledge, this is the most recent comprehensive study of this type that has been conducted in the United States.

Before going into these figures in more depth, however, we need to take a quick look at what goes into these numbers. Both the economic value (sales) and jobs account for direct and indirect impacts. The direct impacts are straightforward. For example, the number of jobs at a nuclear power plant and the monthly payroll would clearly be direct impacts. However, the presence of a nuclear power plant creates additional wealth and jobs in the surrounding community, since the plant workers buy their groceries at the local supermarket, purchase clothing at the local department stores, and send their children to the local schools. All of this creates secondary, or indirect, commerce and jobs. The economic impact often reaches much farther than the local community, since construction materials may be brought in from distant sources—often well outside the state. As the direct jobs grow, new construction is often necessary, including buildings, roads, and so on.

The same multiplying effect occurs for radioisotopes. Direct jobs include the nuclear medical doctors, the food scientists, the airport screeners, the forensic specialists, and so forth. But these people also need a supporting infrastructure—secretaries, accountants, truck drivers, and so on.

The economic model employed for these studies has been used for decades in numerous public and private cases to evaluate a wide variety of economic impacts. Such evaluations are often done to determine what kind of incentives local governments may offer to entice new industries to move into a state or community. Conversely, this approach is often used to determine the impact of closing down a military base. One can likely quibble with some of the fine points of the approach, but the overall merits of the approach are well established. The authors of the present study[3] indicate that they have employed the model in a very conservative fashion, such that they believe the economic impacts of radiation technology are actually higher than reported in their study.

Economic Impacts

Figure 77 contains pie charts to illustrate pictorially the split between the contributions of nuclear energy and radioisotopes to the US economy and the job market. We note that nuclear energy contributes just over 20 percent of the total revenue but only about 10 percent of the total number of jobs.

Figure 78 contains a comparison of the economic contributions of radiation

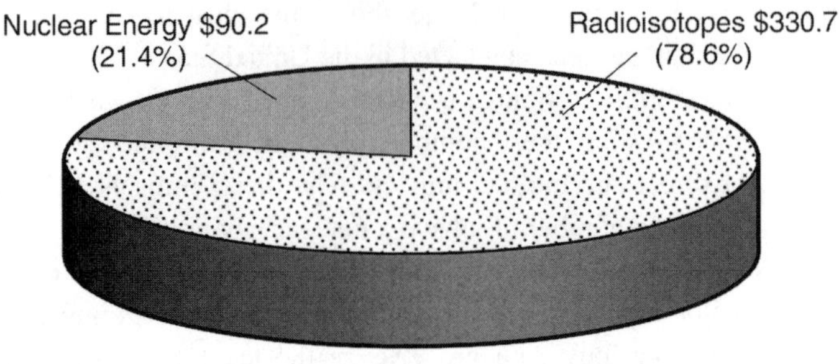

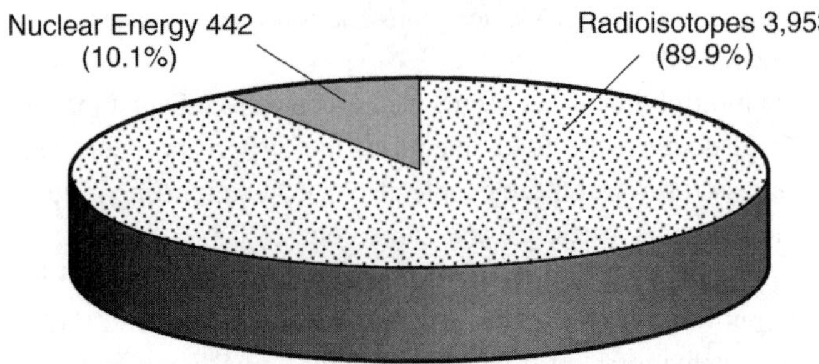

Figure 77. Distribution of economic and job benefits to the US economy by radiation technologies in 1995.

technology and the top Fortune 500 companies in the United States. We note that the total radiation industry of $421 billion is larger than the largest corporation in America; namely, General Motors. It is about the same as Exxon and Wal-Mart combined. Just the radioisotope component alone is larger than either Ford or Chrysler.

Appendix E contains a listing of the economic impact of radiation and nuclear energy (as well as the associated jobs) according to several industry types. The largest impact is on the electric, gas, and sanitary services sector, followed closely by the health, education, and nonprofit sector. As would be expected from our earlier discussions, we note from this listing a wide variety of industries that benefit from both radiation and nuclear energy.

If we compare the economic contributions of radiation technology to the *total* economy of several nations, we are confronted with an even more stunning

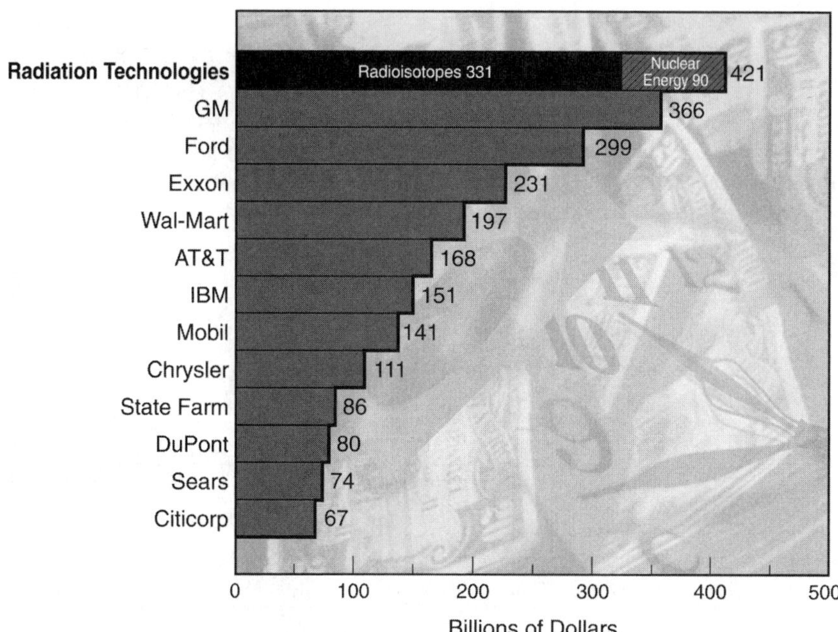

Figure 78. Comparison of the 1995 US economic impact of radiation technologies with that of Fortune 500 companies.

perspective. Figure 79 reveals that the $421 billion value of radiation technologies in the United States equates to 72 percent of the gross domestic product (GDP) of the entire nation of Canada. It is larger than that of Mexico, South Korea, the Netherlands, Belgium, Sweden, and Argentina. In fact, if the economic value of radiation technologies in the United States could be thought of as representing a separate nation, it would be the eleventh largest nation on Earth!

Job Types

The jobs created by radiation technology provide an even more interesting picture. Figure 80 contains a comparison of radiation technologies with several readily identifiable sectors of the US economy. In 1995 radiation technologies created almost as many jobs as the entire banking industry. It created even more jobs than industries such as electronics, printing and publishing, hotels, legal services, and so on. Since I sometimes think I live in airplanes, I find it fascinating that US radiation technologies provide well over three times as many jobs as the total US airline industry.

Another way to gain a perspective on the impact of radiation on the job

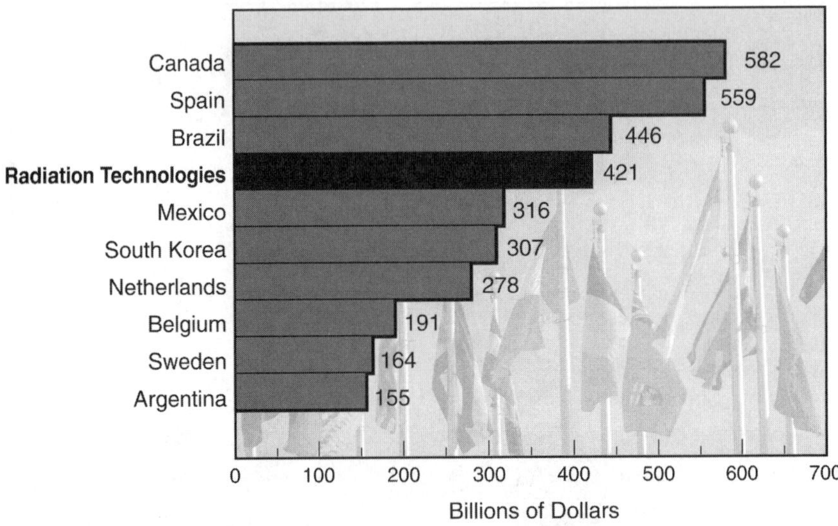

Figure 79. Comparison of the 1995 economic impact of US radiation technologies with the gross domestic products of major countries.

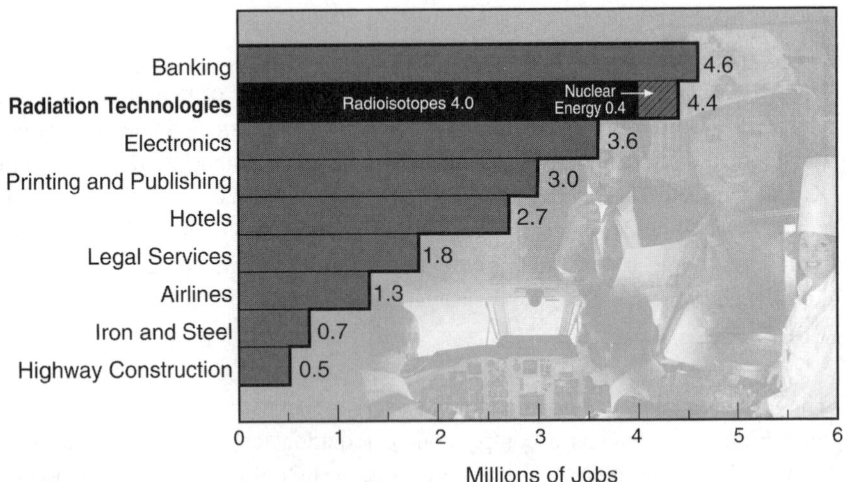

Figure 80. A comparison of the jobs created by radiation technologies with those created by major industries (United States, 1995).

market is to view the number of jobs that radiation technology creates within selected industries. Figure 81 reveals 111,000 jobs in the restaurant industry, 84,000 jobs in finance, 69,000 jobs in the rubber industry, 67,000 hotel jobs, and so forth. The complete listing is contained in appendix F, where we note the creation of over 76,000 truck-driving jobs, approximately 44,000 mechanics jobs for industrial machine repairs, 90,000 jobs in the janitorial and cleaning business, nearly 30,000 receptionists, and so on. The seventy-five different job classifications contained in appendix F clearly reveal the widespread impact of radiation technologies on Main Street America. Many of these jobs are filled at a relatively low-skill level, and at least a million of them are represented by organized labor.

Being an engineer, I was interested in observing the split of engineering disciplines in the radiation industry. Figure 82 contains a pie chart with this information. About one-third of the engineers employed in this field are electrical engineers, followed by mechanical, industrial, and chemical engineers. Nuclear engineers constitute only 3 percent of the total—again illustrating the diversity of the radiation field. On the other hand, the highest percent of total engineering jobs dedicated to the radiation field within each discipline is in nuclear engineering (23 percent), followed by industrial (12 percent), metallurgical (11 percent), electrical (10 percent), and mechanical engineering (9 percent).

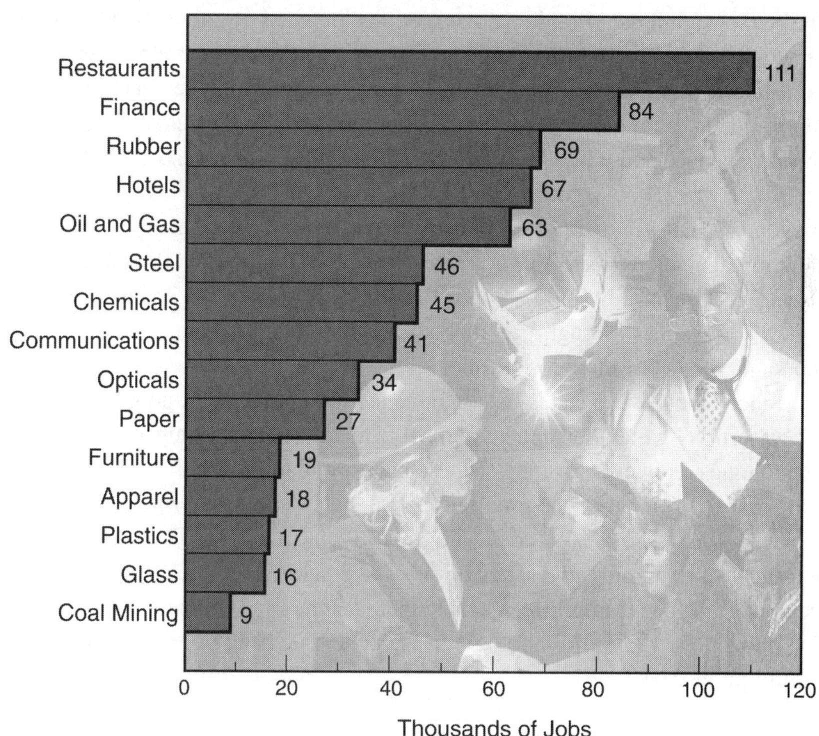

Figure 81. US jobs created in 1995 by radiation technologies within selected industries.

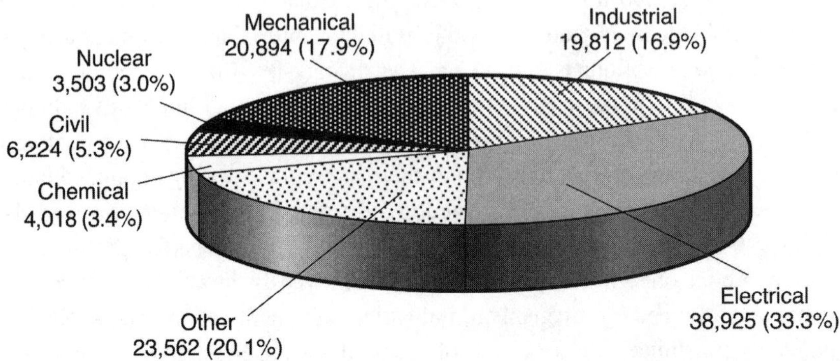

Figure 82. US engineering jobs generated by radiation technologies in 1995.

Impact within the States

The study[4] summarized in the above sections included a state-by-state analysis. As might be expected, the impacts varied quite widely among the states. Appendix G contains the numbers for sales, jobs, and tax revenues by state. Pennsylvania benefited most from 1995 radiation technology, registering $63 billion in revenues and 630,000 jobs. States next in line for revenue benefits were Tennessee ($29B), Virginia ($28B), California ($27B), Texas ($22B), and Oregon ($21B). Following Pennsylvania, the leading states for jobs created by radiation technology were Virginia (379,000), California (250,000), Washington (243,000), Oregon (241,000), South Carolina (241,000), and Texas (220,000). Figure 83 provides a rough visualization of the states benefiting most from the economic impact of radiation technologies.

Since many of the states listed above are among the most populous, it may not be surprising that they benefited most from the economic impact of radiation technologies. However, in four of these states, radiation technology contributed more than 10 percent of the total number of statewide jobs: South Carolina (13 percent), Tennessee (13 percent), Virginia (12 percent), and Pennsylvania (11 percent). An additional five states owed over 5 percent of their jobs to radiation technology (New Hampshire, Connecticut, Minnesota, Oklahoma, and Utah). The impact of radiation technology on the economic health of the United States is truly impressive.

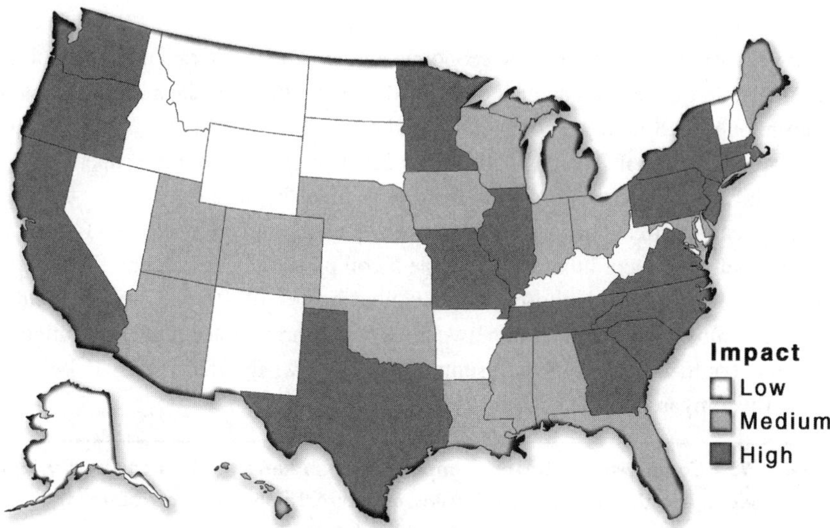

Figure 83. Economic and employment impacts, by state, of the use of radiation technologies in 1995.

Tax Revenues

We note from table 4 that \$79 billion in tax revenues was collected from the US radiation industry in 1995. That averages out to \$1.6 billion per state. Here again the distribution is quite varied, with tax revenues ranging from a low of \$22 million in North Dakota to a high of \$11,231 million (\$11B) in Pennsylvania (see, by way of comparison, appendix G). These numbers should be of special interest to governors and state legislatures.

In particular, we note that in 1995, over \$5 billion was contributed by radiation technology to the California tax coffers. The recent recall election for the California governorship, largely driven by the huge state debt, is but one indication of how important revenue-generating technologies have become to both our modern lifestyle *and* the degree of political stability. No figures are available for economic contributions from the radiation industry to California in 2004, but I would be surprised if they are not a good bit higher than they were in 1995.

JAPAN

Buoyed by the recognition of the economic and job impacts of radiation technology in the United States, researchers conducted similar studies in Japan. In a series of papers published in 2002, comparable studies were recorded for the year 1997. A direct comparison of Japanese and US impacts was covered in these papers.[5]

The Japan investigators chose to report direct impacts only, rather than the multiplicative affects of direct and indirect contributions. Hence, two tables are shown below to record the results. Table 5 compares the direct economic and job impacts of Japan with those of the United States in 1997, whereas table 6 includes multiplicative factors between two and three to account for a range of indirect contributions. This representation also allows the Japan study to be more directly compared to the earlier 1995 US study.

Table 5. A Comparison of financial impacts from radiation technology in 1997 on Japan and the United States (using only direct financial contributions)
All numbers are in $ billions.

		United States	Japan
Radiation	Medicine	49	12
	Agriculture	14	1
	Industry	56	39
	Subtotal	119	52
Nuclear Energy*		39	47
TOTAL		158	99
% GDP		1.9	2.3

* Delivered average KW-hr price assumed: Japan = 15.0 cents; USA = 6.8 cents

Table 6. A comparison of financial impacts ($B) of radiation technology in 1997 on Japan and the United States (using a range of multiplicative factors to account for indirect effects)

	DIRECT		MULTIPLICATIVE FACTOR					
			2.0		2.5		3.0	
	US	JAPAN	US	JAPAN	US	JAPAN	US	JAPAN
RADIATION	119	52	238	104	298	130	357	156
NUCLEAR	39	47	78	94	98	118	117	141
ENERGY*	—	—	—	—	—	—	—	—
TOTAL	158	99	316	198	396	248	474	297
%GDP	1.9	2.3	3.8	4.7	4.8	5.8	5.7	7.0

* Delivered average KW-hr price assumed: Japan = 15.0 cents; USA = 6.8 cents

In comparing the impact of radiation technology on Japan with that of the United States, we must first recall that in 1997 the United States had over twice the population of Japan (270 million to 120 million). Further, the GDP of the United States was about twice that of Japan ($8,318 billion versus $4,231 billion). A value of 121 yen to the dollar was used for the exchange rate.

It is interesting to note that the total economic impact of radiation technology on Japan is actually higher than in the United States, when done on a percentage basis (about 2.3 percent of the GDP in Japan and about 1.9 percent of the GDP in the United States, using direct impacts only). The main difference is that radiation (nonpower) is about three times that of nuclear power in the United States, whereas these two are about equal in magnitude in Japan. The major reasons for this difference in balance are as follows. In 1997 nuclear power constituted about one-third of the total generation of electricity in Japan, whereas it was only about one-fifth of the total generation capacity in the United States. Also, the cost of electricity in Japan is approximately double that in the United States. The application of radiation to the industrial sector[6] is relatively higher in Japan than in the United States, mainly because of Japan's large use of electron beams, x-rays, ion beams, and thermal neutrons in its gigantic semiconductor industry. Japan is known the world over for its strong position in the electronics market, and radiation technology is one of its principal assets in manufacturing electronics products. Japan also employs electron beam accelerators heavily in the vulcanization of rubber for its massive tire industry.

As of 1997, radiation technology was used only sparingly in the field of agriculture in Japan (mainly for crop improvement via mutation), but nuclear medicine is catching on—with about half the applications of the United States in terms of the percentage of its economy.

Depending upon which multiplying factor we choose from table 6 to account for the indirect economic factor, the impact of radiation technology on Japan and the United States in 1997 ranged from about 4 percent to 7 percent of the GDP, with Japan holding a slight overall edge.

CHINA

Studies regarding the impact of radiation technology in China are harder to find. But draft material kindly provided to me by professional colleagues was helpful in allowing us to gain at least a rough assessment of how this new technology is being implemented in the most populous nation on our planet.[7]

Table 7 contains a summary of the available data for China. We note that the bottom-line reported value is $4.3 billion (this utilizes an exchange rate of about 8 Chinese Renminbi, RMB, to the US dollar). Assuming 1997 values, this total corresponds to only about 0.35 percent of the GDP of China (i.e., only about 10 percent of the total national economic impact of radiation technology for either Japan or the United States). However, the numbers provided do not include nuclear medicine (no data were available for that sector) nor nuclear energy. According to the second source in reference 7, some forty-five hundred professionals are working in the nuclear medicine field in China, utilizing some eight hundred machines for nuclear diagnosis work. Nuclear medicine is reaching approximately four hundred million Chinese citizens, and the practice is increasing at the rate of about 10 percent per year. Hence, the economic impact is likely quite significant (even though there are no hard data to provide substantiation).

Table 7. Economic impacts of radiation technology in China for 1997, based on direct financial contributions ($ millions)

INDUSTRY			3,200
	Semiconductors	2,400	
	Cross-Linking	390	
	Nuclear Instrumentation	360	
	Radioisotopes	50	
AGRICULTURE			1,140
	Mutation Breeding	480	
	Food Irradiation	660	
MEDICINE			???
TOTAL*			4,340

* The GDP of China in 1997 was approximately $1,250 billion. Hence, this represents approximately 0.35 percent of China's GDP.

The Chinese nuclear power program did not get tangibly underway until about a decade ago. Two units at Daya Bay in southern China came on-line in 1994 with a total electrical capacity of just under 2,000 MW (about 2 percent of the current US capacity). Hence, the economic impact of nuclear energy would likely have been less than half of the reported values for radiation in 1997. As of this writing, China has added several more nuclear power plants, totaling approximately 6,500 MW of electricity capacity, and they are continuing to be engaged in a strong building program.

Refocusing on table 7, it is noteworthy that electron beam accelerators are often used in China's semiconductor industry. Like Japan, China has carved out a significant foothold in the international electronics market. Substantial effort is going into the manufacture of heat-shrink materials and the use of radiation to cross-link insulation materials for wires and cables. That particular segment of the Chinese industry is reported to have gone up by a factor of ten in the past decade.

Another growing area is the use of electron beams to clean sulfur and nitrogen oxides out of the flue gas from coal-fired power plants. This may become a significant factor in China's attempts to minimize atmospheric pollution. With 1.3 billion people (over 20 percent of the global population) and the world's largest deposits of coal, it is only natural that China's leaders wish to utilize this form of energy to accelerate economic growth. Whereas the country's long-range energy plans are placing substantial emphasis on nuclear power, both to be able to locate energy production parks near the population centers and to minimize pollution, the fact remains that coal will very likely be China's primary energy source for quite some time. Hence, using radiation technology to clean the flue gas could be significant. It won't help ameliorate the problems of global climate change (since it cannot prevent CO_2 from being vented into the atmosphere), but it could minimize the blight of acid rain and "Brown Clouds."

The impact of radiation technology to date on China's agriculture has been primarily the breeding of new crop varieties. Over forty new plant species, including some six hundred varieties, have been recently produced to increase the productivity of China's agriculture. Radioisotopes have been used for numerous tracing applications, as well as for the sterile insect technique (SIT) to eradicate unwanted insect pests. Food irradiation is underway in China, mainly with the use of cobalt-60 as a gamma source. Over a hundred thousand tons of food are currently being irradiated on an annual basis. Given the number of mouths to feed in this crowded nation, food irradiation is very likely to grow in China's future.

Having had the opportunity to visit several of China's nuclear laboratories and academic institutions, I am convinced that Chinese scientists will continue to find new and expanded ways to enable radiation technology to serve their huge and growing population.

Also, having walked a portion of the Great Wall of China—an answer to a childhood dream—and viewed some of the country's spectacular countryside, I am particularly hopeful that China will encourage the widespread use of radiation technology to stem atmospheric pollution and eradicate the "Brown Cloud." This is essential in order to allow the population to grow in prosperity in a manner compatible with environmental stewardship.

Whereas there are yet very little hard data regarding the economic impact of radiation technology for most of the nations of this planet, sufficient information has already been gathered to provide us with a glimpse of its powerful and growing significance. The beneficial impacts of harnessed radiation have not only enriched our medical, agricultural, and industrial services. They have also provided an enormous number of jobs and added considerably to our national economies.

14. A Day with the Atom

What if Marie Curie and Pierre Curie had never discovered that a phenomenon called "radiation" existed? And what if no one else discovered it? One way to gain a better appreciation for the marvels derived from radiation is to go through a typical day, conscious of the role that ionizing radiation plays in nearly every aspect of our lives.

Our alarm clock goes off and the day begins. We wearily look at the clock or our wristwatch, now aware that we can see the dial through the darkness because the dial is luminescent. We then flip the light switch, with full expectations of instant illumination. If we live in the United States, there is a 20 percent probability that the electricity delivered to power those lights comes from a nuclear power plant. In France that probability would rise to nearly 80 percent. Even if the power came from coal, the most probable source worldwide, we know that at least some of the pollutants currently discharged directly into the atmosphere might eventually be removed by electron-beam processing.

We now go to the bathroom and use the toilet. Whereas most sewage is currently treated by chemical means, gamma rays from radioisotopes can be used to process sewage without generating additional waste streams, since the products can be used for fertilizer. The recognition that this technology is on the way resonates with our environmental awareness. We then shower and either insert our contact lenses or affix our eyeglasses. The saline solution in which our contact lenses were stored overnight had been irradiated to kill any microbes that may irritate our tender eyes. If, instead, we choose to put on our eyeglasses, we now recall that several radiation procedures were used to assure a high quality of glass or plastic for this special use.

As we put on our clothes, we appreciate the fact that the cotton in our undergarments is now grown in a more productive fashion, owing to improved cotton strains developed using radiation-mutation breeding procedures. Clothes made from synthetics likely also benefited from radiation cross-linking or other materials-improvement processes somewhere in the development phase. In particular, we select the blue polyester-cotton blended shirt. Radiation was used to bind special chemicals to the fabric to assure a "fresh pressed" look all day. For those who wish to apply cosmetics, you can safely use them (unless you are allergic) because of radiation sterilization processes that rid them of any microbes that could otherwise irritate your skin.

We now trudge into the kitchen and head straight for the refrigerator, thankful that plentiful supplies of electricity allow us to keep our many food products cold and free of excessive spoilage. We have a desire this morning for a fried egg, and we are glad that our egg carton contains eggs with unbroken shells. Eggs with thin, breakable shells were screened out with radiation thickness gauges before they ever got to the grocery store. We proceed to fry an egg on a pan equipped with a special coating to inhibit sticking. Radiation gauges likely determined the thickness of this coating and assured that it was properly adhered to the metal. The ceramic or plastic plate upon which we slide our fried egg could have benefited from radioisotopes to assure uniformity of the materials of the plate itself. Our silverware undoubtedly benefited from radiation thickness gauges both during the making of the sheet metal from which the utensils were subsequently stamped and during the special coating process. For the first time, we become aware that our full packages of breakfast cereal were precisely measured by radiation leveling or density gauges to ensure that the weight and volume stated on the cardboard containers were accurate.

Making one more trip to the refrigerator, we appreciate the array of fruits and vegetables that we have for our selection. Many of these varieties would not have been possible were it not for the greatly accelerated agricultural breeding process made possible by using radioisotopes for either mutation or tracing purposes. Furthermore, the amount of water needed to grow these crops is now substantially less than only a few years ago, because of the use of neutron moisture gauges to determine the minimum amount of water necessary for optimal production. The papaya we are having for breakfast was picked and then irradiated in Hawaii to prevent pests from coming to the mainland. That treatment also considerably extended its freshness. As we pour creamer into our coffee or tea, we marvel that this creamer can remain on our shelf for long periods without refrigeration because the container was irradiated prior to being filled in order to assure the absence of microbes.

As we reach for the morning paper, our mind now flashes back to the paper mills responsible for making such huge amounts of paper available to us so economically. This affordability is largely due to the radiation thickness gauges that allow the paper-production process to be precisely and automatically controlled with amazingly high production speeds. We then flip on the radio or TV, now cognizant that the wiring in these devices is very likely protected with radiation-treated heat-shrink insulation.

After finishing breakfast, we struggle to the medicine cupboard where we take our vitamins and/or our prescribed medication, now aware that such modern marvels would not be possible without the radioisotope tracers employed in so many parts of the development, testing, and FDA approval process. Even the food supplements, conveniently packaged for our use, likely benefited from nutrient tracing techniques afforded by the use of radioisotopes.

Before leaving for work, we prepare a sandwich for lunch, again reflecting on the conveniences afforded by shrink-wrap or aluminum foil that radiation processes helped to produce. We hope that the slices of turkey, ham, or beef placed in our sandwich do not contain *Salmonella*, *Trichinella*, or *E. coli*. It should have been sufficiently cooked to remove these dreaded concerns, but we may look forward to the day when all such foods have been irradiated—to be absolutely sure they are safe.

It's about time to leave for work, but the baby's cry indicates that a quick diaper change may be in order. Thankfully, the super-absorbent material used in the disposable diaper (a direct result of radiation grafting) makes the job easy, without having to change the bedding. We leave our baby in the capable hands of her nanny, who drove up minutes earlier.

As we get into our car to drive to the office, school, or factory, we are thankful that the engine starts quickly and smoothly. Much of the credit for this vast improvement over earlier models is due to advanced materials for the engine—made possible by using radioisotopes to determine engine wear, lubricant levels, and so on. All the steel used in the car benefited directly from radiation techniques, both in the original foundry and in the final metal-rolling process. The tires were likely vulcanized by radiation, rather than by the older sulfurization process. All the glass in the vehicle was perfected in the manufacturing process via radiation moisture monitors. We are thankful for smooth pavement to drive on, now aware that the firmness of the road bed was probably assured via radiation compaction instruments before the final surface was applied.

We roll down the window and breathe in the clean air, a rarity in many places. The use of radioisotopes is becoming more widely used everyday to pinpoint the sources of atmospheric pollution, a necessary step for successful abatement programs. With the advent of nuclear-generated electricity, less coal must be burned. We no longer need to worry that our lungs will be attacked by small particulate pollutants that we may not even be able to see, much less deal with the plight of more-visible smog. Our path to our work location takes us through an agricultural area once fumigated with pernicious chemical sprays to kill unwanted insects. But today the air is fresh, since the SIT radiation process has been used to rid the orchards and fields of those pesky fruit flies.

We finally approach our place of work. If we are fortunate, the floor sparkles as it reveals a wood-grain surface hardened for beauty and providing easy maintenance via radiation-induced cross-linking in the polyethylene materials. We are relieved that all exits in the building are clearly marked by illuminated exit signs, powered by a radioactive source that is 100 percent reliable (even when the electricity goes off). As we approach the drinking fountain, we are grateful that the best water supplies locally available were likely found by using radioactive tracer techniques to assess the suitability of our groundwater hydrology.

Our office experience is especially upbeat this morning because the new furniture order just arrived. We notice that special water-resistant fabric (processed by radiation cross-linking) now covers our chair and a beautiful wooden desk now occupies center stage. We are confident that it will likely never crack since a meticulous drying process was carefully monitored by radiation moisture gauges to prevent subsequent structural damage. During our morning coffee break, a colleague drops by to inform us that the dreaded serial killer was brought to justice when a DNA match, enabled by special radiation techniques, convinced the jury that the suspect was guilty. We also now know that such DNA tests have been used in many cases to prove that a person was not guilty—thus adding to our peace of mind that justice is now fairer.

Having finished a crucial task by noon, we decide to reward ourself with a quick visit to the local museum featuring a traveling masterpiece. We know that such a painting will be authentic, since any forgery would have been previously discovered using radiation dating processes.

Once back at the office, our afternoon drowsiness can be remedied by a quick trip to the vending machine, where we can find a wide selection of canned beverages—all filled to the advertised volume by radiation level gauges. We've been waiting for that all-important e-mail message and there it is. Our e-mail and other computer work now function with enormous capacity, largely because of the incredible power that microcircuitry (using radiation-enhanced semiconductors) can deliver in our highly compact work stations.

After work we recall our friend who's been recuperating in the hospital and decide to pay a visit. Our new awareness of nuclear medical advances reminds us that one out of every three patients entering such a facility derives direct benefits from radiation. Our friend is doing nicely because his ailment was detected quickly and precisely using radioactive diagnostic methodology. The PET scan revealed an abnormality that could be treated in a conventional fashion, rather than through unnecessary surgery. Given our friend's situation, we decide to stop by the radiology department and schedule our annual chest x-ray. We are also reminded that dental care is now much better because of the routine diagnostic x-rays taken prior to performing actual corrective measures. As we leave the hospital, we stroll past the special care unit, now acutely aware that many of these patients will recover, because of new cures for cancer and heart disease becoming available from the use of specialty radioisotopes.

We then return home for dinner and automatically turn up the heat or the air conditioner. Again, the electricity delivered for such service could be from a nuclear power plant. Our dinner is seasoned with spices that almost certainly have been irradiated to prevent insect infestation. A good juicy steak can be eaten without concern if it has been irradiated prior to shipment. An alternate choice is that leftover turkey, which we could take out of the refrigerator and pop into the microwave oven (another device that depends upon a form of radiation to perform its essential function). When the turkey was purchased, it was covered with an irradiated polyethylene heat-shrink wrap—a special convenience for assuring an airtight seal around the meat. We then top off our meal with a large scoop of our favorite ice cream, having become aware that the constituency of this treat was enhanced by the use of radiation devices to assure the correct mix of air at a crucial time in the manufacturing process.

After dinner we buzz over to the airport to pick up our favorite uncle. We are thankful that the lights illuminating the runway are likely powered by tritium, a radioactive substance that continues to operate independent of any electrical failures or storm conditions. We are equally grateful that the welds in the wings of the airplanes are routinely inspected using neutrons or gamma rays from special radioactive sources, as are the welds affixing the jet engines to the wings. Likewise, all luggage boarding the aircraft is screened using radiation procedures to minimize the threat of concealed weapons or explosive devices. As a result of these advances, air travel, despite our concerns, is considerably safer today than any other mode of transportation.

We arrive home just in time to turn on the TV for the late-evening news, and relish the footage of the landing of our latest spacecraft on one of Jupiter's moons. This event would not have been possible without radioisotope power systems that originated in the work of Marie Curie and those dedicated scientists who followed her lead.

Finally, it is time to climb into bed and get a good night's sleep. Such contentment is aided by the knowledge that our trusty smoke detector, which operates with a built-in radioisotope, is 100 percent reliable throughout the day and night.

As we reflect back over the day, we should be no less than astounded by the degree to which radiation processes have enriched our life. Recognizing this enormous progress, made largely over the past half century, we can only dream in wonderment over what the future of radiation technology may hold for us and for our children.

I was once impressed by the slogan "A day without radiation is a day without sunshine." Whereas this statement is certainly true, I feel confident that we all now recognize that we receive a good deal more from radiation on a daily basis than just those rays that emanate from the center of our solar system.

15. A GLIMPSE INTO THE FUTURE

Having described some of the myriad applications of harnessed radiation we already enjoy, I will now venture to speculate on how many more ways radiation might be able to serve humanity over the next century. What types of life-transforming applications can we reasonably be expected to encounter? They will likely be mind-boggling by today's standards.

But before we attempt such speculation, we might be reminded of where all of this got started. Over a century ago, Marie Curie spent countless days and nights in poorly lit quarters, shivering through the cold Paris winters while stirring vat after vat of a uranium slurry to recover trace quantities of what she had earlier discovered to be radioactive daughter products of uranium-238—ones that she had named polonium and radium. Ironically, on the day of this writing, National Public Radio (NPR) reminded me in its newscast that today (December 21, 2003) is the 105th anniversary of those original discoveries. How in the world could Marie Curie have possibly imagined how such elusive materials could one day lead to numerous new agricultural plant varieties, sophisticated medical diagnosis techniques, actual cures for some types of cancer, revolutionized industrial processes, a tool for solving crimes, protection of passengers at airports, a probing of outer space some billions of miles from Earth, and even the ability to calculate how long ago our planet was formed?

For us to be so bold as to try imagining what new applications might come into existence over the next hundred years will likely be equally challenging. The only thing I know for sure is that there will be new discoveries in this time frame that will spawn directions we can't even dream of at the present time. That is the power of science and discovery. Hence, I suspect that we will likely undershoot—

and probably undershoot badly. When our great-grandchildren pick up this book a hundred years from now and blow off its dust, they will likely laugh until their sides hurt—wondering how our imagination could have been so limited.

AGRICULTURE

Growing more and more food on less and less tillable farmable acreage will become increasingly more challenging as we plunge deeper into the present century. With the global population at least 50 percent higher, and possibly doubling by the year 2100, pressure will continue to mount in order to find ways to harvest greater yields. Given their track record to date, there is every expectation that radiation mutation techniques will continue to be refined and employed to develop even more robust and productive varieties of grains, grasses, vegetables, and fruits to provide essential nutrition. Irrigation water will almost assuredly become a premium in many parts of the world, requiring energy from nuclear power plants to produce potable water from the oceans. This source of consumable water is very likely to become commonplace as a principal means to grow essential produce, as well as to supply other water needs of sprawling metropolitan areas.

There are still vast areas of the globe where farmers burn the stubble in their fields as a measure to rid their land of insects and other pests that thrive on commercial plant life. With concerns for atmospheric pollution continuing to rise, we will likely see burning fields being replaced with farm machines equipped with radioisotope gamma emitters that can be wheeled over large tracts of acreage to sanitize the soil. Much of the fertilizer needed to supply essential nutrients could well be derived from the local sewage-disposal plant, where gamma sources are used to kill unwanted bacteria and provide recycled biomass.

Perhaps the biggest contribution of radiation to agriculture over the next century will be the irradiation of our food. Citizens of the world will simply not have the luxury of allowing high percentages of field produce to spoil. Furthermore, the public will demand that foods purchased at the local supermarket be sanitized—both to prevent spoilage and to be assured that unwanted pathogens and bacteria have been removed. With the food-irradiation infrastructure fully in place, the added cost of food irradiation could very well be more than offset by the savings afforded by avoiding food spoilage.

MEDICINE

It is simply impossible to predict what nuclear medicine may be like in another century. The advances made even within the past decade are sufficiently dazzling that to make an attempt to envision year 2100 seems like pure folly. Perhaps the best way to approach this topic is to envision what we really want to happen in advanced health programs. My guess is that those in the nuclear medicine community will play a strong role in helping to find ways to achieve those wishes.

Like most people, I look forward to the day that diagnostic and therapeutic techniques are available to analyze and treat *my* particular needs. As impressive as medicine is today, most patients are treated with drugs in a generic fashion—using a statistical approach to prescribe drugs that work in *most* situations. But with radioactive tracers, I can foresee the day when specialty radioisotopes can be attached to unique chemical carriers capable of tracing the human anatomy to discern maladies unique to any individual. This would enable the attending physician to prescribe medicine to treat very specific, individualized problems. In all probability, it will be a combination of biochemical advances, along with gene therapy, and nuclear medicine that will be used in a collaborative process to diagnose and treat not only life-threatening but also chronic diseases.

One of the most frustrating and difficult maladies that we have to deal with today is mental disease. Anxiety and depression are far more prevalent in today's society than most will care to admit. Despite the relatively comfortable lifestyle that modern technology has afforded most of us in the developed world, the stresses have become enormous. I seriously doubt that there is a family in America today that has not been at least indirectly touched by a member or close relative who has suffered some type of depression. The frustration is that the medical profession still lacks the diagnostic tools necessary to pinpoint the particular drugs needed to get the delicate biochemical balance back in order. It is still pretty much of a "let's try this" approach until finally something works—if indeed it does. The human brain is an exceptionally complex organ, and no two people react to corrective drugs in precisely the same way. I fully expect that advanced radioactive tracer–scanning techniques will eventually be able to determine, with far greater precision than is possible today, how minute metabolic functions take place in the brain—thus allowing medications to be individualized in bringing prompt relief to the millions of people who suffer from this and other dehabilitating maladies.

Brain cancer is still one of the most difficult types of cancer to cure. We mentioned some of the advances underway earlier in this book, but we fully expect much more progress to be made in the decades ahead. Scientists at Brookhaven National Laboratory on Long Island are already probing ways to utilize microbeam radiation therapy for the type of brain tumors that are currently impossible to treat.[1] Some forms of brachytherapy (direct implantation of radioisotopes) or proton-beam therapy may also be perfected for treating brain cancer.

Numerous studies of a very fundamental nature are taking place at the European Synchrotron Radiation Facility[2] in France to probe the mysteries of the human body, including brain functions, possible treatments for Parkinson's disease, asthma, blood clotting, and so on. Given advances in the arteriovenous malformation (AVM) treatment method that we discussed briefly in the chapter on medicine, it may one day be possible to apply an advanced form of this treatment for detecting and curing aneurysms.

It is now becoming possible to employ radiation to temporarily immobilize bioactive materials such as drugs and hormones on polymers. This property is already being used in some new drug-delivery systems, including an eye insert that releases medicine to combat glaucoma and an implant that controls the release of prostaglandin for the treatment of ulcers. The use of irradiated polymers as part of a "smart bullet" package could revolutionize the treatment of diabetes for millions of sufferers.[3]

Whereas organ transplants became the medical miracle in the last half century, the problem remains that of finding suitable donors. Is it possible that we or our children will see the day when our own diseased organs can be physically removed from our body, treated with some combination of radiation and gene therapy, and then reinserted? Only time will tell, but the fact that some pioneering work along these lines is already underway may give us hope that this dream might be realized.[4]

It has often been said that the only sure things in life are death and taxes, but the "fountain of youth" is something that many of us crave. Certainly we all are destined to leave this life at some point, but there is substantial research among those in the medical community focused on how the aging process works. Because of such research, there is now discussion about the possibility of developing microenergy sources that could once again "switch" on DNA cells that our body has turned off, thus extending life itself. It is certainly conceivable that radioisotopes could provide critical links for such possibilities. Radiation is also likely to play a significant role in leveraging off a new understanding of sequencing the human genome.

Many patients today are still phobic about radiation and in some cases will even reject nuclear medical procedures on this basis (not recognizing that the benefits far outweigh the risks). Yet there are medical practioners today in some parts of the world (notably Japan) in which cancer patients are being subjected to rather substantial doses of whole-body radiation—and the cure rates are quite impressive.[5] This approach is totally opposite to the beliefs in the world in which I grew up. I was constantly reminded that we should seek to keep radiation levels as low as reasonably achievable (the ALARA principle) for public protection, and being cautious certainly has substantial merit. But the theory of hormesis (i.e., that low levels of toxic substances actually stimulate the body's immune system) is gaining momentum in many fields, and I can envision the day when nuclear medicine practitioners may prescribe increased levels of whole-body radiation for effective therapy purposes.

Electricity

Access to abundant and affordable supplies of energy is crucial in driving the economy and national defense systems of any country. Whereas this commodity is needed in the form of electricity, heat, and kinetic energy, the segment growing the fastest in essentially all parts of the world is electricity. A modern economy simply cannot function without large and affordable supplies of electricity. Our world in the year 2100 will very likely be consuming at least five times as much electricity as it does today. We noted earlier that there are very few sources capable of such a prodigious supply—especially when we demand that such generating sources be compatible with environmental stewardship. Hence, it is almost certain that nuclear energy will be one of the primary sources of electricity, if not *the* principal one, within the next hundred years.

This kind of a demand will require new types of reactors, ones that squeeze far more energy out of the raw stock of mined uranium. Present-day reactors utilize less than 1 percent of the uranium dug out of the ground (since the fissionable isotope, U-235, exists in nature at only 0.7 percent of the total uranium, and the isotope U-238 is not fissionable in today's reactors). However, it is technically possible to build reactors that can transform energetically inert uranium-238 into plutonium isotopes that are fissionable. They are called breeder reactors.[6] Some concern has been expressed that these devices might produce more plutonium than can be readily burned as fuel, thus posing the possibility of this material falling into questionable hands and being used for illicit purposes. However, it is practical to design these advanced reactors to either produce or consume fissile fuel, thereby allowing a commodity such as plutonium to be carefully balanced and tightly controlled. Furthermore, several such reactors have already been built and successfully operated in nations such as the United States, France, the United Kingdom, Japan, and Russia. China is well on its way to building a demonstration breeder reactor.

These types of reactors have been shown to work dependably over many years of service, but the early versions were quite expensive. Hence, they will not likely be built for commercial purposes for several more decades when they are really needed to extend nuclear fuel supplies. But unless fusion reactors should prove to be both possible and feasible, there is really no other option in the nuclear reactor world. There are large amounts of thorium in some parts of the world, and some scientists argue that this fuel might eventually replace uranium in a mature breeder reactor economy.

Fusion may ultimately work, but I seriously doubt that these devices will contribute more than a very small fraction of commercially viable electricity by the turn of the next century. The technical challenges are enormous. At the present time, there is a good deal of international cooperation to spread the costs of keeping fusion energy research going. The ITER (International Thermonuclear Experimental Reactor) project is the first large international effort to see if fusion power could become a reality.[7] Several nations in Europe, in addition to Japan, Canada, China, Korea, Russia, and the United States, are the key partners. This is a $5 billion project that, if successful, could be the first such machine in the world to create more usable fusion energy than the energy input required to heat materials hot enough to ignite the fusion process.

Figure 84 provides a vision of how nuclear energy may grow to serve an energy-starved world.[8] Today, nuclear power is mostly relegated to the developed nations—or at least those well on the way toward substantial development. If and when the really populous nations of the world fully develop nuclear power as a prime source of their electricity, the dynamics could change vastly.*

Whereas the concern regarding what to do with nuclear waste seems to have a near-paralyzing effect on many energy planners today, I fully expect this issue to be satisfactorily resolved within the next few decades. Once the public recognizes that there are numerous techniques to safely handle and dispense with such relatively small amounts of waste, and the infrastructure is fully in place to implement such practices, I am confident this will become a nonissue. In fact, I am reasonably optimistic that within the next century, we will find ways to utilize beneficially most of what we now call nuclear "waste."

My reason for this is that I was once asked to provide a technical paper on the pros and cons of closing the nuclear fuel cycle (i.e., recycling used fuel) at a Gordon Research Conference. Few people outside the scientific community have ever heard of the Gordon Conferences, because they are conducted on an invitation-only basis for preeminent world scientists on very specialized topics. They are

---

* Figure 84 also includes a possible low future for nuclear energy. It is unclear to me how the citizens of the world could even hope to survive if the low scenario should become a reality.

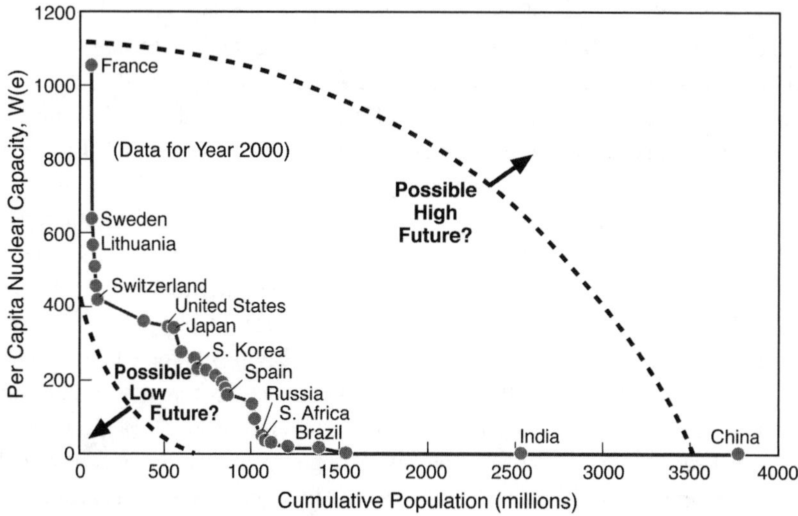

Figure 84. Possible ranges for the development of nuclear energy capacity over the next century.

unique in that no observers are allowed and nothing is reported outside of the conferences themselves without the authors' permission. This may seem strange, but the fundamental purpose is to provide scientists with the cover necessary to discuss their issues in a completely open process with their peers without any concern for political or economic implications. Scientists are, therefore, free to speak their minds without the possibility that their research funding might be cut because of any "political incorrectness." Hence, the verbal crossfire is prolific and intense.

Given this setting, I was almost a basket case when I was called upon to give my paper. This is because I chose to challenge my colleagues by asking why nuclear waste could not be considered a resource for beneficial use, rather than treating it as a dastardly "witches brew" to be rid of as fast as possible. I was certain that I would be the laughing stock of the conference. I suggested, for instance, that two of the most troublesome fission products in spent nuclear fuel, strontium-90 and cesium-137, might be chemically removed from the discarded fuel and put to use. Strontium-90 is a beta emitter with a thirty-year half-life. We know, for example, that a fiber-optic cable laid under the sea from New York to Paris would need energy boosters along the way to keep any communication signals strong. If this energy were to be supplied in a conventional manner, over three thousand miles of copper wire would be necessary. On the other hand, Sr-90 could be embedded in an RTG-type device, similar to those that currently supply electricity for our spacecraft. Plutonium-238 is used in the space program because it is an alpha emitter, thereby requiring very little shielding, which is heavy. But at the bottom of the ocean, shielding is certainly no problem, and a beta emitter like Sr-90 should work quite well. I calculated that the amount of Sr-90 needed for such a purpose would match reasonably well with the amount being discharged from our current fleet of nuclear power reactors. Cesium-137, on the other hand, is a gamma emitter (again with a thirty-year half-life). Why not use it for irradiating municipal sewage, or possibly for food irradiation?

I noted about halfway through my presentation that the facial expressions of my audience gradually mellowed from initial skepticism to actual agreement. To my utter amazement, my colleagues appreciated my courage and even went as far as electing me the chairman of the follow-on conference! Stephen Covey, the best-selling author of *The 7 Habits*, is right; we need to begin with the end in mind.[9]

Other forms of energy will likewise benefit enormously from advanced nuclear technology. By the year 2100, coal will still be a significant source of electricity on the global scale, and electron-beam processes might be used to remove most of the sulfur- and nitrogen-oxidized contaminants out of the flue gas. As noted earlier, the primary limitation to coal-fired plants will be the release of carbon dioxide. The numerous sequestration projects currently underway will need to be highly successful if coal is to remain a viable source of electricity. This means forcing the CO_2 into large underground caverns or depleted oil wells, effectively depositing it in the ocean or dispensing with it in some other as yet untried measure. No other fossil fuel sources (oil or natural gas) are projected to be available by the year 2100 to make anything more than spot contributions to the electricity-generating market. The raw supplies simply will not be available in the amounts required—if available at all.

Solar energy will probably make a substantially stronger contribution to the generation of heat and electricity, and nuclear technology may help to improve the efficiency of solar's electrical contribution. Some studies already indicate that special applications of targeted radiation can cause point defects in solar cells, thereby providing considerably longer life in the photovoltaic (PV) method of converting sunlight into electricity. Both wind and biomass energy will also likely be used to greater degrees in the future, since there will be enormous pressure to derive electricity from any reasonably economic sources.

INDUSTRY

As with medicine, it is hard to imagine what type of products will be available in daily commerce a hundred years from now. I vividly recall going to a technical conference in Las Vegas about ten years ago and noticing the look of disbelief by a Russian colleague who had just arrived in the hotel lobby from Obninsk, a city just south of Moscow. Having been in Obninsk only a few months earlier myself, I could readily understand how the contrast must have almost blinded him. The baffling expression on his face quite likely mirrored the one I first flashed in the early 1960s upon stepping out on the sidewalks of Grand Central Station in New York City. With my neck craned upward, I'm sure I typified the little country kid who had never before seen a real skyscraper and was sure he had arrived in a world inhabited by men from outer space.

One of the major historical contributions to advanced manufacturing systems has been materials development. The invention of iron and bronze marked a major turning point in history. So did the invention of cement. In the past century, plastics have probably had the most transforming power. But what will we find over the next several decades? I expect one thing will be materials that can withstand much higher temperatures. Various ceramic composites will likely be out in front, since substantial work is already being done. But with methods of radiation delivery being continually upgraded, what currently dazzles us such as radiation cross-linking of special chemicals will likely pale into insignificance to the materials that radiation will help synthesize over the new century.

As horrible as war is, history has clearly revealed that one of the beneficial side effects is the rapid development of materials to ward off disaster. "Self-healing" materials research is already well underway for items such as bulletproof chest protectors, and it is not much of a stretch to expect these materials to be commonplace within a few decades. Very efficient gas masks, utilizing specially developed materials, are likely to be commonplace to provide protection from biological and chemical threats.

Superconductivity provides amazing possibilities for a wide range of applications. The major drawback at present is the need to take materials to exceptionally low temperatures before they exhibit the properties necessary to carry large amounts of electricity with low resistance. But progress is inching along to develop materials that exhibit such characteristics at higher and higher temperatures. It is possible that radiation processes will one day allow the development of superconducting materials that could operate in the range of room temperatures. Should that become a reality, it would literally transform our industrial infrastructure. Long-distance trunk lines to carry massive amounts of electricity from coast-to-coast could be envisioned with the realization of such technology.[10]

By now we are all aware of how powerful the world of miniaturization (nanotechnology) has become. There is now far more computational power within a wristwatch computer than existed in the most powerful computers of only a few decades ago. Radiation detection devices are likewise being manufactured in smaller and smaller sizes, which at the same time have become even more sensitive. This means that ever-smaller amounts of radiation will be necessary to do their jobs as tracers in the agricultural, medical, and industrial worlds. Such advances are almost certain to markedly improve the myriad of operations now in common usage—thereby increasing productivity by substantial margins.

TRANSPORTATION

We noted earlier that one of the biggest challenges for continuing to fuel the transportation industry is that a very high percentage of the fuel used to date is petroleum. No matter how we cut it, this raw source will simply not continue to be available within a few decades. Not at any price. Hence, there is enormous pressure to find another transportable fuel.

Electricity will undoubtedly be used more and more in the transportation sector. The electrification of the rail system is already largely complete in many nations, such as Switzerland. We know that nuclear energy can play a major role in powering large ships (based on our experience with nuclear-powered submarines, aircraft carriers, and ice breakers). This practice will almost assuredly grow and will likely contribute considerably to the commercial sector. But how do we fuel trucks and cars in the future?

The answer is most likely to be hydrogen. We introduced this thought in chapter 8. At the present time, however, huge obstacles remain. How do we safely and efficiently store hydrogen? How can it be generated in large enough quantities to really replace oil and gas? Whereas these challenges are large, and much work needs to be done, I personally believe that hydrogen will be the prime source of fuel for our transportation system by the year 2100—and the bulk of it will very likely be generated using the energy from a massive nuclear power infrastructure. It is the only source that can provide hydrogen in the quantities needed without simultaneously contributing to environmental pollution.

There will also be significant advances in providing lighting sources for essential operations. Both tritium and krypton-85 are, as noted, currently being used in illuminating airport runways. But other radioactive sources will likely be developed to make such applications even better. For example, thulium-170 can create a much brighter light than either tritium or Kr-85. It is currently quite expensive, but that could change in a fully developed nuclear infrastructure.

SPACE EXPLORATION

We noted in chapter 9 that well over 99 percent of outer space is simply not explorable by manned spacecraft without nuclear energy. Given the inquisitiveness of the human mind, the space exploration program currently underway could well be compared to Henry Ford's Model T by the time we leap into the next century. But that can happen only by a substantially increased application of nuclear technology.

Heat and power for small spacecraft will likely continue to come from radioisotopes, but the efficiency of transforming the heat energy into electricity will be considerably higher—allowing larger spacecraft to probe significantly farther distances. The much larger demands for electricity and propulsion power required to explore the outer planets and interstellar space will almost assuredly come from nuclear reactors. The source of heat and electricity to support astronaut life on foreign planets and moons will likewise come from fission reactors.

One of the most exciting prospects for advanced space power is the use of antimatter. We were introduced to one form of antimatter earlier when we discussed the use of positrons as a key ingredient in the functioning of the PET scan medical diagnostic device. Recall that a positron is an electron with a positive, rather than a negative, charge. When a positron comes into contact with an electron, both particles literally annihilate and their mass is transferred into energy.

In searching for an advanced fuel for space travel, scientists have been giving serious consideration to another form of antimatter; namely, antiprotons. An antiproton is essentially a proton, but with a negative, rather than a positive, charge. When an antiproton collides with a proton, the two disappear and are transformed into a large amount of pure energy. To appreciate the energy potential of antimatter, it is useful to compare fuels on a per-unit-mass basis. If we arbitrarily assign chemical energy a value of one, fission energy is about ten million times higher, fusion energy is about thirty million times higher, and antimatter energy is about ten billion times higher. Stated in a different way, antimatter reactors have an energy value about a thousand times higher than fission energy and ten billion times higher than the fuel of the *Saturn V* booster rockets.

The principal challenge in harnessing antimatter reactions is to manufacture the fuel itself. With today's technology, antimatter costs about $62.5 trillion to produce a gram (or $1.75 quadrillion an ounce).[11] As such, it is by far the most expensive substance on Earth. It is made in huge accelerators by accelerating protons to very high speeds and then crashing the protons onto a target, usually tungsten. Among other reaction products, a few antiprotons are created. But even at full production, a super accelerator, like the one operating at the Fermi National Accelerator Laboratory in the United States, can produce only about ten billionths of a gram per year! This is far less than that needed for any practical application.

The next obstacle to overcome is to capture these elusive rascals. Since they spontaneously annihilate when brought into contact with normal matter, they must be contained by an electromagnetic field in a very high vacuum. But progress is being made both in the production and storage areas, such that this may prove to be a viable power source for space travel by the turn of the next century. The lure of tapping this nuclear source of power is that it could potentially enable manned missions to the outer planets.

Several designs are underway to utilize scarce commodities of antimatter to leverage the operation of fission/fusion rocket engines. By seeding fissionable uranium with a small quantity of antimatter to produce heat, the number of neutrons ejected from a fission process could jump from two or three to possibly sixteen. This, in turn, could heat a deuterium/tritium fuel to high enough temperatures to ignite a fusion reaction. Although too inefficient to be used as a viable energy source on Earth, this could well be used in a nuclear rocket. If we can let our imaginations soar a bit, we might foresee antimatter becoming the nuclear fuel of the future to conquer distant planets.

TERRORISM, CRIME, AND PUBLIC SAFETY

Whereas it would be romantic to envision a world where crime and terrorism did not exist, the fact is that until all nations can develop an economy that removes dire pockets of poverty, the dark side of humanity will always be present. Even if economic deprivation did not serve as the underlying reason for desperate actions, it is naive to believe that crime will ever vanish. Hence, it will be necessary to continue developing advanced technology to thwart crime and terrorism.

One of the tools that advanced nuclear technology will likely contribute is a system capable of generating exceptionally high bursts of neutrons or other nuclear particles that can neutralize any chemical, biological, or nuclear weapon without itself causing more than localized damage. It is almost certain that there will come a day when a terrorist group will announce to the world that it has in its possession a weapon of mass destruction ready to unleash at any moment. Without a device that can disable such a weapon, the ransom that could be extracted by such a terrorist group could be huge, if indeed they wanted one. The need to develop countermeasures to provide protection will only increase with time.

Other systems that will almost assuredly be commonplace within the next century are "smart" clothing systems whereby our armed forces or even civilians can instantly detect the presence of ambushes or horrible weapons and transmit that information in a manner to bring counterforces to bear. All of this will require small energy sources that could be supplied by radioisotopes, along with very sophisticated portable detection devices.

Arts and Sciences

As significant as the discovery of radiation was in the late 1890s by Marie Curie and her team, even they could not have conceptualized that nuclear fission and nuclear fusion were even remote possibilities. By extrapolation, even though modern physics has unraveled many of the fundamental forces of nature, we really know very little of how the universe actually works. In fact, high-energy nuclear physicists tell us that we understand only about 10 percent of the fundamental forces of nature. More than 80 percent of the universe is made up of what we now call "dark matter." If and when the forces that hold all of this together are understood and can be harnessed, the power that we currently covet as being so impressive may, within a hundred years, be regarded as the equivalent of a small water pistol. Our earlier discussion about exploiting antimatter may provide us with a brief glimpse of what awaits us.

Given the almost-certain development of miniature power sources, it is conceivable that "art-in-motion" could become a reality. With radioisotopes having half-lives of several decades, it is conceivable that new-age artists will find ways to introduce motion into their masterpieces that would seem unbelievable in today's world.

Environmental Protection

It is almost certain that by the year 2100 we will be sweltering and in dire need of water. Since over 80 percent of the globe is covered with oceans, the logical source of our water by that time will be the oceans themselves. Our challenge will be to get the salt out in some reasonable fashion. The deployment of nuclear reactors to desalinate seawater could very well be one of the largest uses of nuclear power by the year 2100.

Even prior to such a massive development, it may become feasible to utilize radiation on a much smaller scale to purify water in global regions during emergencies. We are often startled to hear of either natural or man-made disasters that leave villages or entire cities and their surroundings in dire need of emergency food and water. I know of at least one physicist who believes that contaminated water could be irradiated and converted into safe drinking water. Part of the cleansing could come from direct electron collisions with bacteria and pathogens, but some could also come from the cleansing effects of free radicals created in the water from beta bombardment. Wouldn't it be marvelous to be able to purify water for such disasters right on site?

PERSPECTIVE

The above thoughts may seem to paint a future picture far too unrealistic for some readers. To others, it may seem to be severely lacking in imagination. The famed New York Yankee catcher Yogi Berra was once quoted tongue-in-cheek to remark that one can predict pretty much anything—except the future. How true!

Yet short of some cataclysmic event that wipes out the human imagination, I remain confident that radiation will continue to make important—perhaps even astounding—contributions to the world of our children and their children's children. Figure 85 reveals that well over half of all the elements contained in the periodic table contain radioisotopes that have already been harnessed in some fashion to serve a plethora of beneficial humanitarian applications. These include medicine, industry, environmental protection, and others such as agriculture, electricity, transportation, space exploration, public safety, and the arts and sciences. It will be most illuminating to revisit this chart in another hundred years and note the additional advances.

Given the vast array of innovations we already enjoy that derive from radiation, it is difficult to overstate the beneficial contributions that Marie Curie and those who followed her have made to modern living.

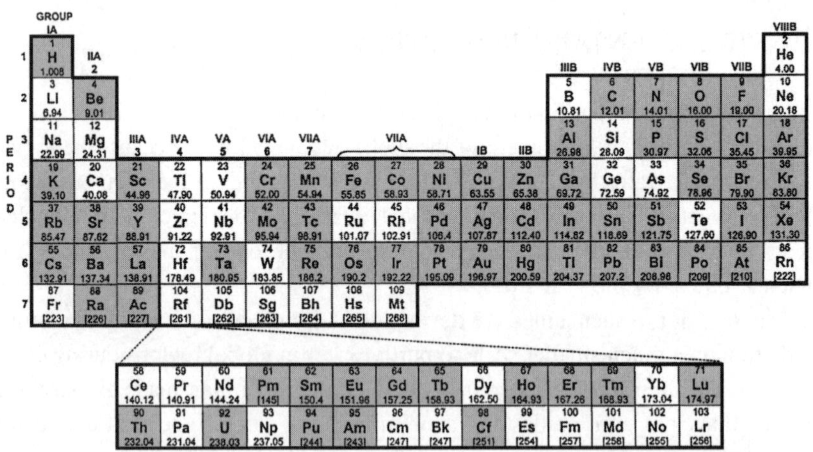

Figure 85. Well over half of all elements (shaded boxes) in the periodic table contain radioisotopes that have already been harnessed to serve a plethora of humanitarian uses in everyday living.

Epilogue

In reflecting back over the pages of this book, I find it no less than startling to appreciate how many areas of our everyday lives have been beneficially affected by harnessed radiation. Over the past decade or so I have had the opportunity to become generally aware of the myriad beneficial impacts of radiation. But in the process of writing this book I continually learned of many more. Yet I know that I have been able to only scratch the surface. I fully expect many astute readers to point out numerous other areas of application.

One of the great pleasures of writing this book was to work directly with Hélène Langevin-Joliot. She not only prepared a unique and inspiring introduction, but she also reviewed the entire manuscript and made numerous valuable corrections. Perhaps her major contribution was to remind me continually that science and technology, as important they are, will not in and of themselves solve the monumental problems that face human progress. Rather, it is how we use such new technological insights.

We need to be reminded that science is ethically neutral. Good science drives new discovery and new understanding. Good engineering builds upon such discoveries to create new applications for beneficial use. But in the end, it is the direction of the human spirit and imagination that controls how these discoveries and subsequent products are put to use. Some people fear technology, not recognizing how important new scientific discoveries are to the progress of human civilization. Some will even exploit fear—playing on the emotions of unsuspecting citizens to achieve their own hidden agendas to gain recognition or power. Certainly there are cases in which technological breakthroughs and their potential applications should be challenged. And clearly the nuclear industry should be monitored. But technological advances bring enormous progress when handled with intelligence and discretion. The harnessing of ionizing radiation is an example of where the benefits far outweigh the risks.

One of the greatest challenges we face in this new century is to find ways to generate energy in sufficient and affordable quantities to light and heat our homes and factories, provide potable water, fuel our transportation needs, grow and safely deliver adequate food supplies, and provide the medical aid needed for good health—all consistent with environmental stewardship. Fossil sources can help for several more decades, but at the price of huge volumes of contamination being discharged into our atmosphere, and the geopolitical issues associated with securing rapidly depleting resources. Solar and wind energy will also help, but they will require a dedication of large land masses and backup energy sources because of their intermittent availability. Nuclear energy can also help, but at the expense of sequestering small amounts of waste that require diligent control. We need to muster the courage and dedication to balance these options, in full recognition that to turn our backs on any of them is likely the biggest risk of all.

In addition to holding the promise of long-term, environmentally compatible energy supplies, we now know that radiation technology is providing enormous benefits in the fields of agriculture, medicine, industry, transportation, space exploration, public safety, the arts and sciences, and environmental protection. Based on this progress to date, we can look forward to even more ways to employ radioisotopes to serve humanity in the future. The rate of progress is largely up to us. Both individually and collectively, we can choose to either support or oppose continued research to capture and harness the monumental potential of radiation technology.

Marie Curie set an excellent example for us to follow—both with her creative scientific mind and with her unbending devotion to utilizing discoveries for the greatest benefit to human progress. It is our challenge and opportunity to help fulfill her ultimate dream—and ours.

APPENDIXES

Appendix A. Radioisotopes Used in Medicine*

Element	Radioisotope	Primary Decade Mode**	Half-life	Major Applications
Actinium	Ac-225	alpha	10.0 days	Monoclonal antibody attachment for cancer therapy; decays ultimately to Bi-213 (monoclonal antibody attachment) with four alpha particles in the decay chain
	Ac-227	alpha	21.8 years	Decays to Ra-223 (monoclonal antibody and radium chloride treatment of bone cancer)
Astatine	At-211	alpha	7.21 hours	Monoclonal antibody attachment for cancer therapy
Barium	Ba-133m	gamma	1.62 days	Calibration source for radiation dose measurements
Bismuth	Bi-212	alpha	1.01 hours	Monoclonal antibody attachment for cancer therapy (leukemia and solid-tumor metastases)
	Bi-213	alpha	45.6 minutes	Monoclonal antibody attachment for cancer therapy (leukemia and solid-tumor metastases)
Californium	Cf-252	alpha, SF (neutrons)	2.65 years	Brain cancer therapy
Carbon	C-11	positron	20.4 minutes	PET imaging; measure physiological processes in the brain and other organs
	C-14	beta	5,715 years	Radiolabeling for detection of tumors (breast); drug metabolism studies
Cesium	Cs-131	x-ray	9.69 days	Brachytherapy for prostate seed cancer therapy
	Cs-137	gamma	30.1 years	Blood irradiation to make organ transplants successful
Chromium	Cr-51	gamma	27.7 days	Cell labeling and dosimetry; blood volume studies

Appendix A. Radioisotopes Used in Medicine* (*continued*)

Element	Radioisotope	Primary Decade Mode**	Half-life	Major Applications
Cobalt	Co-57	gamma	272 days	Gamma camera calibration; source for x-ray fluorescence spectroscopy
	Co-60	gamma	5.27 years	Sterilization of medical equipment; polymerization of plastics used in prostheses; blood irradiation to make organ transplant successful; external beam radiation cancer therapy
Copper	Cu-64	positron	12.7 hours	PET imaging; SPECT imaging; monoclonal antibody therapy of cancer
	Cu-67	beta	2.58 days	Cancer therapy (breast); monoclonal antibody therapy of cancer
Erbium	Er-169	beta	9.40 days	Arthritis treatment; cancer therapy
Europium	Eu-152	gamma	13.5 years	External beam radiation cancer therapy
	Eu-155	beta	4.75 years	Osteoporosis detection
Fluorine	F-18	positron	110 minutes	PET imaging; measuring glucose activity in the brain; functional heart imaging
Gadolinium	Gd-153	gamma	242 days	Osteoporosis detection
Gallium	Ga-67	gamma	78.2 hours	Imaging of abdominal infections; detect Hodgkins and non-Hodgkins lymphoma; osteoporosis detection
	Ga-68	positron	67.8 minutes	PET imaging; detection of pancreatic cancer
Gold	Au-198	beta	2.70 days	Cancer therapy
	Au-199	beta	3.14 days	Cancer therapy
Holmium	Ho-166	beta	1.12 days	Bone marrow treatment, liver and multiple myeloma cancer therapy; arthritis treatment
Indium	In-111	gamma	2.80 days	Label white blood cells for infection; imaging label for Y-90 monoclonal antibodies; therapeutic agent for radiolabeled peptides and antibodies

Iodine	I-123	gamma	13.2 hours	Brain, thyroid, kidney, and myocardial imaging radiography in molecular biology
	I-124	positron	4.18 days	PET imaging
	I-125	x-ray	59.4 days	Brachytherapy for prostate seed cancer therapy; eye cancer; diagnostic imaging: radiography and protein labeling in molecular biology
	I-131	beta	8.02 days	Thyroid cancer therapy; radiolabeled monoclonal antibody for therapy of breast, colon, and lymphoma cancer
Iron	Fe-55	positron	2.73 years	Tracer in studies of iron metabolism and blood perfusion
	Fe-59	beta	44.5 days	Tracer in studies of iron metabolism and blood perfusion; therapy of liver cancer
Iridium	Ir-192	beta	73.8 days	Brachytherapy for various cancer therapy (brain, breast, gynecological, lung, head and neck, tongue, and mouth)
	Ir-194	beta	19.3 hours	Cancer therapy
Lutetium	Lu-177	beta	6.65 days	Radiography monoclonal antibody for cancer therapy (Hodgkin's disease, leukemia, breast, liver); heart disease (restenosis); rheumatoid arthritis therapy
Molybdenum	Mo-99	beta	66.0 hours	Decays to Tc-99m (the most important radioisotope used in nuclear medicine for a multitude of diagnostic imaging studies)
Nitrogen	N-13	positron	9.97 minutes	PET imaging
Osmium	Os-191	beta	15.4 days	Decays to Ir-191m (six-second half-life), which provides gamma rays for cardiac studies (particularly pediatric)
Oxygen	O-15	positron	2.04 minutes	PET imaging
Palladium	Pd-103	x-ray	17.0 days	Brachytherapy for prostate seed cancer therapy

Appendix A. Radioisotopes Used in Medicine* (*continued*)

Element	Radioisotope	Primary Decade Mode**	Half-life	Major Applications
Phosphorus	P-32	beta	14.3 days	Ovarian cancer and leukemia therapy; treatment of multiple myeloma and polycythemia rubra vera; biomedical applications (DNA analysis, protein analysis, and labeling)
	P-33	beta	25.3 days	Biomedical applications (DNA analysis, protein analysis, and labeling)
Platinum	Pt-195m	Auger electron	4.01 days	Monoclonal antibody attachment for cancer therapy
Radium	Ra-223	alpha	11.44 days	Monoclonal antibody attachment for cancer therapy; chloride form for treating painful skeletal metastases from breast and prostate cancer
	Ra-224	alpha	3.66 days	Generator (target nuclide) for Bi-212
	Ra-226	alpha	1,599 years	Target material (parent nuclide) for producing several alpha-emitting radioisotopes used in cancer therapy
Rhenium	Re-186	beta	3.72 days	Inhibits reclosure of arteries after angioplasty procedure; bone cancer pain relief; monoclonal antibody therapy of cancer
	Re-188	beta	17.0 hours	Monoclonal antibody attachment for cancer therapy; pain relief for bone cancer, heart disease (restenosis); bone marrow ablation; liver cancer therapy; arthritis treatment
Rubidium	Rb-82	positron	1.26 minutes	PET imaging; early detection of coronary artery disease; myocardial imaging
Samarium	Sm-153	beta	1.93 days	Pain relief for bone cancer
Scandium	Sc-47	beta	3.35 days	Pain relief for bone cancer
Strontium	Sr-89	beta	50.52 days	Pain relief for bone cancer
Sulfur	S-35	beta	87.2 days	Biomedical applications (DNA analysis, protein analysis, and labeling)

Technetium	Tc-99m	gamma	6.01 hours	Multitude of nuclear medicine diagnostic imaging studies, and for identifying cancer and skeletal mestastases; tracing technique to determine bodily disease (brain, heart, liver, gallbladder functions, spleen, lungs, bones, thyroid, parathyroid, kidney, red and white blood cells)
Thallium	Tl-201	x-ray	3.04 days	SPECT imaging; heart imaging; tumor imaging; treadmill tests
Thorium	Th-228	alpha	1.91 years	Decays to Bi-212 (monoclonal antibody attachment)
	Th-229	alpha	7,300 years	Decays to Ac-225 and eventually to Bi-213 (monoclonal antibodies)
Tin	Sn-117m	electron capture	13.6 days	Pain relief for bone cancer
Tritium	H-3	beta	12.3 years	Drug metabolism studies
Xenon	Xe-127	beta, gamma	36.4 days	Neuroimaging for brain disorders
	Xe-133	beta, gamma	5.24 days	Lung and liver imaging; SPECT imaging of brain; measure regional cerebral blood flow
Yttrium	Y-90	beta	2.67 days	Cancer therapy and as labeled monoclonal antibodies for treating Hodgkins disease, non-Hodgkins lymphoma, breast, liver, and colon cancer; as microspheres for treating primary liver cancer, treating benign diseases (rheumatoid arthritis); treating restenosis of clogged arteries in heart disease

*The primary modes of decay and half-lives (rounded to three significant figures) for the radioisotopes contained in appendixes A through D were taken from the Chart of the Nuclides, Knolls Atomic Power Laboratory, 16th edition, 2002. We note that many of these radioisotopes have several modes of decay (some occurring simultaneously). For the present tables, we have generally included only the mode of decay most appropriate to the applications cited.

**Whereas the term "primary decay mode" is normally descriptive, it may be better to use the term "primary energy released" in some cases, such as the release of neutrons or x-rays. For instance, x-rays are the manifestation of electron capture (absorption of an inner-shell electron by the nucleus). Neutrons are generally released in conjunction with spontaneous fission (SP). Also, the word "primary" is used to distinguish the principal decay of most interest. In some cases, other decay modes are statistically more prevalent.

NOTE: The author is indebted to the following professionals who scrutinized and contributed significantly to the compilation of appendixes A through D: Dr. Robert Lull, Dr. Darrell Fisher, Dr. Robert Schenter, Dr. Manuel Lagunas-Solar, and Dr. Peter Airey.

Appendix B. Radioisotopes Used in Industry

Element	Radioisotope	Primary Decade Mode	Half-life	Major Applications
Americium	Am-241	alpha, gamma	433 years	Smoke detectors; combines with beryllium to produce neutrons for material inspections (aircraft and others), oil well borehole analysis, soil density measurements, and coal ash measurements; used with targets (e.g., copper or silver) to generate pure fluorescent x-ray sources
Antimony	Sb-124	beta	60.2 days	With beryllium and heavy water, produces neutrons for variety of applications
Cadmium	Cd-109	gamma	461 days	Calibration of detectors for XRF (x-ray fluorescence)
Californium	Cf-252	alpha, SF (neutrons)	2.65 years	Spontaneous fission neutrons for conducting materials radiography in aircraft and other crucial structures; soil density measurements
Carbon	C-14	beta	5,715 years	Thickness gauges for thin coatings
Cesium	Cs-137	gamma	30.1 years	Determine ash content of coal entering a power plant (with Am-241); determine soil density; radiation monitoring calibrations
Cobalt	Co-57	gamma	272 days	Mossbauer spectrographs
	Co-60	gamma	5.27 years	Radiation hardness testing of electronics used in space applications; radiation testing of components and materials for nuclear power plants; sterilization of products; blast furnace performance; heat-shrink plastics; determine soil density; tracing water around oil wells; nondestructive examination (radiography)
Gold	Au-198	gamma	2.70 days	Blast furnace performance; aluminum electro-refining performance

Iridium	Ir-192	gamma	73.8 days	Nondestructive examination (radiography)
Iron	Fe-55	x-ray	2.73 years	Low-energy x-ray analyses
Krypton	Kr-85	beta	10.8 years	Thickness gauges; airport runway lights; methane gas tracer
Lanthanum	La-140	beta	1.68 days	Blast furnace performance; heat-shrink plastics
Lead	Pb-210	gamma	22.3 years	Energy standard for XRF (x-ray fluorescence)
Mercury	Hg-197m	gamma	23.8 hours	Trace mercury impurities
	Hg-203	gamma	46.6 days	Trace mercury impurities
Nickel	Ni-63	beta	101 years	Measure thin coatings on thick substructures; small power sources
Plutonium	Pu-238	alpha	87.7 years	Low-energy XRF (x-ray fluorescence) measurements
	Pu-239	alpha	24,100 years	With beryllium, neutron production for weld inspections
Promethium	Pm-147	gamma	2.62 years	Thickness gauges
		beta	2.62 years	With aluminum, produces x-rays to trace water or methane around petroleum fields
Selenium	Se-75	gamma	120 days	Nondestructive examination (radiography)
Strontium	Sr-90	beta	28.8 years	With aluminum, produces x-rays to trace water or methane around petroleum fields; thickness gauges; small power sources
Tantalum	Ta-179	x-ray	1.82 years	x-ray flourescence
Terbium	Tb-160	beta, gamma	72.3 days	Nondestructive examination (radiography)
Thallium	Tl-204	beta	3.78 years	Measure thin coatings on thick substructures
Thulium	Tm-170	gamma	129 days	Low-energy radiography
Tritium	H-3	beta	12.3 years	Airport runway lights; with zirconium target, x-rays for tracing water or methane round petroleum fields; luminescent exit lights

Appendix C. Radioisotopes Used in Environmental Protection

Element	Radioisotope	Primary Decade Mode	Half-life	Major Applications
Aluminum	Al-26	gamma	7.1×10^5 years	Measure sediment accumulation over past few million years (combination with Be-10)
Argon	Ar-39	beta	269 years	Dating groundwater supplies
Beryllium	Be-7	gamma	53.3 days	Measure rates of sediment accumulation
	Be-10	beta	1.5×10^6 years	Measure sediment accumulation over past few million years
Bromine	Br-82	gamma	1.47 days	Measure water flow
Carbon	C-14	beta	5,715 years	Dating of groundwater supplies, glacial ice fields, and coral fields up to fifty thousand years old; study CO_2 uptake in oceans; study mixing processes in upper layers of oceans; dating of sediment patterns in reservoirs and estuaries
Cesium	Cs-137	gamma	30.1 years	Determine soil erosion or sedimentation patterns in reservoirs or estuaries
Chlorine	Cl-36	beta	3.01×10^5 years	Dating of groundwater supplies up to a few million years; study salinity processes in groundwater
Chromium	Cr-51	gamma	27.7 days	Measure erosion patterns for up to four months
Copper	Cu-64	gamma	12.7 hours	Measure ecological impact of heavy-metal contamination
Gold	Au-198	gamma	2.70 days	Measure plumes of pollutants in bodies of water; measure transport of sand and sediments over about one week; evaluate ocean sewage outfalls; determine optimal location of dumping sites for dredge spoil
Iodine	I-125	gamma	59.4 days	Labeling tracer to detect growth hormone abnormalities in shellfish
Iridium	Ir-192	gamma	73.8 days	Trace sand and sediment dispersion patterns for about six months

Lanthanum	La-140	beta	1.70 days	Measure short-term erosion patterns (few days)
Lead	Pb-210	gamma	22.3 years	Measure sediment accumulation in reservoirs or estuaries (up to one hundred years)
Manganese	Mn-54	gamma	312 days	Measure ecological impact of heavy-metal contamination
Nickel	Ni-63	beta	101 years	Pollution measurement instruments
Phosphorus	P-32	beta	14.3 days	Measure ecological impact of phosphate contamination
Promethium	Pm-147	beta	2.62 years	Pollution measurement instruments
Radium	Ra-224	alpha	3.66 days	Tracer material to determine origin of air masses
	Ra-226	alpha	1,599 years	Used in combination with Th-220 to validate radioisotope transport models for nuclear waste in geologic strata
Scandium	Sc-46	gamma	83.8 days	Measure erosion patterns over about nine months; determine optimal location of dumping sites for dredge spoil
Silver	Ag-110m	gamma	250 days	Measure erosion patterns over about one year
Technetium	Tc-99m	gamma	6.01 hours	Evaluate ocean sewage outfalls; measure dynamics of inland sewage ponds, water flow, and air contamination
Thorium	Th-230	alpha	75,400 years	Measure long-term sediment accumulation; used in combination with U-234 to measure transport of solutes and particles in ocean currents
	Th-232	alpha	1.4×10^{10} years	Dating of geologic material up to a billion years or older
Tritium	H-3	beta	12.3 years	Measure water flows, plumes of pollutants in bodies of water, ocean surface mixing patterns, and dynamics of inland sewage ponds; identify groundwater recharge areas

Appendix C. Radioisotopes Used in Environmental Protection *(continued)*

Element	Radioisotope	Primary Decade Mode	Half-life	Major Applications
Uranium	U-234	alpha	2.46×10^5 years	Used in combination with Th-230 to measure long-term sediment patterns in reservoirs or estuaries; measure natural dispersion of nuclear waste in geologic strata
	U-235	alpha	7.04×10^8 years	Used in combination with stable lead-206 to date geologic material up to about one billion years old; used in combination with U-238 to measure natural dispersion of nuclear waste in geologic strata
	U-238	alpha	4.47×10^9 years	Used in combination with stable lead-206 to date geologic material up to over a billion years old
Zinc	Zn-65	gamma	244 days	Measure ecological impact of heavy-metal contamination

Appendix D. Radioisotopes Used in Other Applications

Element	Radioisotope	Primary Decade Mode	Half-life	Major Applications
Aluminum	Al-26	gamma	7.1×10^5 years	**Arts and Sciences**: Dating archeological and geologic objects of about a million years old
Americium	Am-241	alpha	433 years	**Public Safety**: Smoke detectors; neutron production (with beryllium) for airport luggage screening
Argon	Ar-39	beta	269 years	**Arts and Sciences**: Dating crystalline structural materials
Barium	Ba-133m	gamma	1.62 days	**Public Safety**: Materials screening at ports of entry
Beryllium	Be-10	beta	1.5×10^6 years	**Arts and Sciences**: Dating artifacts a few million years old
Californium	Cf-252	alpha, SF (neutrons)	2.65 years	**Transportation:** Inspection of materials (airplanes)
Carbon	C-14	beta	5,715 years	**Agriculture:** Tracer for nutrients in animal's digestive system **Arts and Sciences**: Dating archeological, geological objects, and groundwater up to about fifty thousand years old
Cesium	Cs-137	gamma	30.1 years	**Agriculture:** Food irradiation **Public Safety**: Cargo screening at ports of entry **Future:** Irradiation of food, irradiation of municipal sewage
Cobalt	Co-60	gamma	5.27 years	**Agriculture:** Food irradiation
Krypton	Kr-85	beta	10.8 years	**Transportation:** Airport runway lights
Nickel	Ni-63	beta	101 years	**Public Safety:** Detection of explosives and drugs
Phosphorus	P-32	beta	14.3 days	**Public Safety:** DNA analysis

Appendix D. Radioisotopes Used in Other Applications (contiued)

Element	Radioisotope	Primary Decade Mode	Half-life	Major Applications
Plutonium	Pu-238	alpha	87.7 years	**Space Exploration:** Heat for spacecraft essentials; heat for producing electricity for spacecraft
	Pu-239	alpha (neutrons)	24,100 years	**Electricity:** Fissile material that readily fissions to produce heat for generation of electricity in nuclear power plants
	Pu-241	alpha (neutrons)	14.4 years	**Electricity:** Fissile material that readily fissions to produce heat for generation of electricity in nuclear power plants
Polonium	Po-210	alpha	138 days	**Space Exploration:** Heat for spacecraft essentials; heat for producing electricity for spacecraft **Public Safety:** Reduce static electricity in paper-duplicating operations
Potassium	K-40	gamma	1.27×10^9 years	**Arts and Sciences:** Dating crystalline structural materials of approximately a billion years old
Tritium	H-3	beta	12.3 years	**Transportation:** Airport runway lights **Public Safety:** Exit sign illumination
Strontium	Sr-90	beta	28.8 years	**Future:** Power source for underwater fiber-optic cables
Uranium	U-235 (neutrons)	alpha	7.04×10^8 years	**Electricity:** Fissile material that readily fissions to produce heat for generation of electricity **Transportation:** Propulsion power for submarines, aircraft carriers, and icebreakers
	U-238	alpha	4.47×10^9 years	**Electricity:** Fertile material that can be transmuted into Pu-239 for generation of electricity in power plants **Arts and Sciences:** Used in combination with stable lead-206 to date geologic objects of approximately a billion years old; defense ordinance; shielding for radiographic sources

Appendix E: Sales and Jobs Created in the United States in 1995 within Selected Industries by Nuclear Energy and Radioisotope Technologies

Industry Title	Sales (million $)			Jobs		
	Energy	Radioisotopes	Total	Energy	Radioisotopes	Total
Livestock and livestock products	53	1,753	1,806	366	15,221	15,587
Iron ore mining	23	981	1,004	103	5,532	5,635
Maintenance and repair construction	4,742	5,062	9,804	36,199	48,633	84,83
Lumber and wood products, etc., containers	233	2,531	2,764	1,989	27,211	29,200
Paper and allied products	243	4,825	5,067	1,029	25,795	26,825
Chemicals and selected chemical products	756	9,491	10,247	2,673	42,277	44,950
Rubber and miscellaneous plastics products	303	6,802	7,105	2,343	66,213	68,555
Primary iron and steel manufacturing	495	7,118	7,614	2,390	43,239	45,630
Heating, fabricated metal products	289	1,534	1,823	2,562	17,119	19,682
Engines and turbines	623	1,028	1,651	2,761	5,747	8,508
General industrial machinery	180	1,840	2,020	1,299	16,727	18,026
Electrical transmission equipment	255	3,322	3,577	2,179	35,754	37,933
Transportation and warehousing	3,615	10,915	14,530	34,536	131,300	165,836
Electric, gas, and sanitary services	40,756	13,620	54,376	118,586	49,918	168,505
Wholesale and retail trade	1,428	11,705	13,133	23,910	246,888	270,798
Finance and insurance	1,437	5,695	7,131	14,096	70,384	84,480
Hotels and personal services	204	2,373	2,577	4,290	62,758	67,048
Business services	1,903	15,819	17,722	24,181	253,116	277,297
Health, educational and nonprofit	147	47,546	47,603	2,657	1,081,805	1,084,462
Total all industries*	$90,151	$330,739	$420,890	442,406	3,953,461	4,395,866

*Total includes all industries not listed separately. Source: Management Information Services, Inc., 1996.

Appendix F: Jobs Generated Nationwide within Selected Occupations in 1995 by Nuclear Energy and Radioisotope Technologies

Occupation	Jobs	Occupation	Jobs
Financial managers	14,156	Heavy equipment mechanics	13,127
Purchasing managers	7,607	Industrial machinery repairers	43,916
Managers, medicine and health	6,234	Data processing equipment repairers	2,412
Accountants and auditors	41,020	Heating and air conditioning mechanics	7,824
Management analysts	3,655	Mechanical controls repairers	3,729
Personnel and training specialists	14,073	Millwrights	3,924
Inspectors, except construction	7,257	Brickmasons and stonemasons	2,824
Architects	2,914	Glaziers	968
Electrical engineers	38,925	Structural metal workers	1,136
Industrial engineers	19,812	Supervisors, extractive occupations	11,993
Surveyors and mapping	835	Explosives workers	1,633
Chemists	6,595	Mining machine operators	5,454
Geologists and geodesists	6,921	Tool and die makers	14,412
Veterinarians	1,756	Precision grinders	995
Pharmacists	4,542	Sheet metal workers	7,338
Economists	3,291	Upholsterers	1,897
Technical writers	3,407	Optical goods workers	2,375
Photographers	3,827	Inspectors and testers	10,956
Clinical laboratory technicians	17,971	Water and sewage plant operators	3,456
Drafting occupations	26,095	Drilling machine operators	5,817
Chemical technicians	7,394	Forging machine operators	679
Computer systems analysts	16,374	Metal plating machine operators	9,484
Tool programmers	95	Sawing machine operators	4,917
Legal assistants	3,957	Photoengravers and lithographers	1,649
Sales representatives	45,647	Textile sewing machine operators	16,978
Supervisors, financial records	3,816	Packaging/filling machine operators	17,999
Receptionists	29,568	Separating machine operators	3,760
Personnel clerks	4,388	Crunching/grinding machine operators	2,166
File clerks	11,441	Photo process machine operators	2,939

Appendix F: Jobs Generated Nationwide within Selected Occupations in 1995 by Nuclear Energy and Radioisotope Technologies (*continued*)

Occupation	Jobs	Occupation	Jobs
		Welders and cutters	25,145
Dispatchers	6,991		
Weighers and checkers	3,759	Solderers and brazers	9,888
General office clerks	26,137	Graders and sorters	2,598
Proofreaders	599	Truck drivers, heavy	76,556
Statistical clerks	5,302	Parking lot attendants	1,207
Supervisors, guards	1,181	Operating engineers	7,086
Kitchen workers	3,976	Machine feeders and offbearers	7,237
Janitors and cleaners	90,003	Vehicles and equipment cleaners	7,229
Transportation attendants	2,616		
		Total, all occupations*	4,380,497

*Includes jobs not listed here. Source: Management Information Services, Inc., 1996

Appendix G: Total Jobs, Sales, and Tax Revenues Created within Each State in 1995 by the Use of Nuclear Technologies

State	Sales ($ thousand)	Jobs	Tax Revenues ($ thousand)
Alabama	$4,844	42,735	$894
Alaska	307	3,599	71
Arizona	3,539	26,144	658
Arkansas	1,855	13,314	346
California	26,536	250,117	5,157
Colorado	3,169	37,853	569
Connecticut	12,248	113,662	2,730
Delaware	795	8,092	145
District of Columbia	428	2,002	97
Florida	5,212	38,313	973
Georgia	9,672	94,285	1,758
Hawaii	1,556	23,931	288
Idaho	335	4,041	61
Illinois	15,558	124,528	2,861
Indiana	3,692	41,119	664
Iowa	4,132	64,055	745
Kansas	1,534	12,466	284
Kentucky	1,641	18,257	299
Louisiana	2,666	21,428	513
Maine	2,143	29,863	384
Maryland	6,963	77,480	1,458
Massachusetts	11,110	110,010	2,306
Michigan	4,799	37,880	892
Minnesota	12,961	132,149	2,413
Mississippi	1,837	16,161	339
Missouri	7,240	78,000	1,300
Montana	351	4,570	66
Nebraska	1,775	14,594	328
Nevada	945	10,502	192
New Hampshire	924	4,970	202
New Jersey	14,667	126,686	3,057
New Mexico	546	5,823	103
New York	18,501	147,841	4,010
North Carolina	12,432	128,846	2,251
North Dakota	112	997	22
Ohio	6,460	68,960	1,222
Oklahoma	6,413	72,030	1,141
Oregon	20,761	241,381	3,834

Appendix G: Total Jobs, Sales, and Tax Revenues Created within Each State in 1995 by the Use of Nuclear Technologies *(continued)*

State	Sales ($ thousand)	Jobs	Tax Revenues ($ thousand)
Pennsylvania	62,901	629,616	11,231
Rhode Island	291	3,245	54
South Carolina	19,382	240,990	3,493
South Dakota	327	3,607	60
Tennessee	29,173	235,766	5,194
Texas	21,834	220,456	3,919
Utah	3,233	41,110	598
Vermont	1,299	11,793	248
Virginia	28,246	379,137	5,042
Washington	18,421	243,381	3,277
West Virginia	973	11,255	178
Wisconsin	3,818	32,573	706
Wyoming	335	4,255	65
Total	$420,890	4,395,867	$78,700

Source: Management Information Services, Inc., 1996

Glossary

absorption. The retention, in a material, of energy removed from radiation passing through it.

accelerator. A device that accelerates electrically charged atomic or subatomic particles to high energies.

accelerator mass spectrometer. An instrument used to determine the masses of atoms or molecules.

actinides. Elements with atomic number 90 through 103.

activation analysis. A method to identify an unknown material by irradiating it and measuring the characteristic radiation from the unknown material.

activity. Short for radioactivity. The number of disintegrations per unit time taking place in a radioactive material.

ALARA. As low as reasonably achievable (a principle to keep radiation levels as low as practical).

alpha particle. A positively charged particle emitted by certain radioactive material, made up of two neutrons and two protons (a helium nucleus).

AMTEC. Alkaline metal thermal to electric conversion. A device that converts thermal energy into electricity using liquid metal ions.

annihilation radiation. The electromagnetic radiation resulting from the mutual annihilation of the mass of negatively charged electrons or antiprotons and positively charged positrons or protons.

antimatter. Material consisting of atoms that are composed of positrons, antiprotons, or antineutrons.

arc-jet engine. An electromagnetic propulsion engine used to supply motive power for flight; hydrogen and ammonia are often used as the propellant, and some plasma is formed as the result of electric-arc heating.

atom. The basic component of all matter. It is the smallest part of an element having all the chemical properties of that element. An atom consists of a nucleus (that contains protons and neutrons) and surrounding electrons.

atomic bomb. A highly explosive device deriving its energy from nuclear fission.

Atoms for Peace. An initiative of President Eisenhower in December 1953 to allow the peaceful uses of atomic energy to be available to other nations.

attenuation. A reduction in radiation intensity upon passage through matter.

background radiation. The ionizing radiation in the environment to which everyone is exposed.

backscattering. The deflection of radiation or nuclear particles by scattering processes through angles greater than ninety degrees with respect to the original direction of travel.

beta particle. A negatively charged particle (a high-speed electron) emitted in certain types of radioactive decay.

BNCT. Boron neutron capture therapy; a medical procedure in which boron is injected into the patient and it subsequently concentrates at the site of a malignancy. A neutron beam is then focused on the malignancy and alpha particles are emitted from the boron to kill the cancer.

borehole logging. The placement of radiation sources into a well to determine the potential for finding economically viable deposits of oil.

brachytherapy. A radiation medical treatment that places a solid or enclosed radioisotope source inside the body directly into or in close proximity to the area being treated.

breeder reactors. Nuclear reactors that generate more useable fuel than they consume.

bremsstrahlung. Electromagnetic radiation that is emitted by an electron accelerated in its collision with the nucleus of an atom. The radiation is visible as a beautiful blue hue.

BWR. Boiling water reactor; a light water-cooled nuclear reactor in which some boiling occurs.

CANDU. Canada Deuterium Uranium reactor; a nuclear reactor cooled by deuterium (heavy water).

carbon dating. A technique using carbon-14 to determine the age of geologic materials or historical artifacts.

CAT scan. Computed axial tomography; a device capable of forming a three-dimensional image of the body by computationally combing numerous two-dimensional x-ray images.

chain reaction. A continuing series of nuclear fission events that take place inside a nuclear reactor. Neutrons produced in one fission event cause another fission event.

Chernobyl. A nuclear reactor in the former Soviet Union (Ukraine) that suffered a major accident in 1986 when technicians tried to perform an experiment with all safety systems disabled.

cladding. Structural tubing that encases nuclear fuel.

Compton scattering. The elastic scattering of photons by electrons.

computerized tomography (CT). The process used to conduct a CAT scan.

conformal radiotherapy. A medical procedure in which x-rays or protons are directed into a patient from many angles in a precise manner, with the beam always focused on the tumor. However, the radiation only penetrates healthy tissue for a short period of time. This is sometimes called the "gamma knife."

containment system. A heavy structure completely surrounding a nuclear reactor to prevent radioactivity from getting into the atmosphere in the event of a major accident.

coolant. A liquid or gas used in a nuclear reactor to remove the heat generated by the fission process.

cooling tower. A towerlike device in which atmospheric air circulates and cools warm water in the outer coolant loop of a power plant.

core. The center fuel region of a nuclear reactor.

cosmic rays. Naturally occurring radiation that exists in outer space (radiation coming from the cosmos). Many of these rays are absorbed by the atmosphere.

cyclotron. A machine to accelerate charged atomic particles to very high energies in a circular fashion by the application of electromagnetic forces.

dating. The use of radioisotopes to determine the age of geologic materials or historical artifacts.

decay. The disintegration of the nucleus of an unstable nuclide by spontaneous emission of charged particles, photons, or both.

deuterium. A nonradioactive isotope of hydrogen having one proton and one neutron in its nucleus. When combined with oxygen, it is called "heavy water."

DIPS. Dynamic isotope power system; a piston-driven system used as an engine to generate electricity for space vehicles.

dose. An amount of radiation or energy absorbed (often measured in mrem).

e-beam processing. A radiation device to remove sulfur and nitrogen oxides from the flue gas of a coal-fired power plant.

electron. A basic particle that has a negative electrical charge and has very little mass compared to a proton or neutron.

electron microscope. A microscope utilizing electrons, rather than visible light, to magnify molecular-scale objects.

electron volt (eV). A unit of energy equal to the energy acquired by an electron when it passes through a potential difference of one volt in a vacuum.

element. An atom with a unique number of protons in its nucleus. Oxygen is an element that has eight protons in its nucleus.

fast neutrons. Commonly defined as neutrons of energy exceeding about 10 keV.

fertile. A material that becomes fissile upon absorbing a neutron.

fissile. A material that will fission, that is, split into two or more lighter materials, upon absorbing a neutron.

fission. The process of splitting a heavy atom into two or more lighter atoms upon absorbing a neutron. This process generates a large amount of energy and usually at least two neutrons.

fission products. Elements that remain after nuclear fission has taken place.

fuel. Material that can be converted into useful energy.

fuel cycle. All the steps required to supply, use, and process fuel for nuclear reactors, including the disposal of nuclear wastes.

fusion. Fusing two light atoms into one heavier atom (a process that releases an enormous amount of energy).

gamma ray. A high-energy, short-wavelength electromagnetic radiation emitted in the radioactive decay of many nuclides. Gamma rays originate from the nucleus.

geologic repository. A burial ground deep beneath the earth's surface designed for the long-term storage of high-level nuclear waste.

glove box. A sealed and shielded box with gloves attached and passing through opening into the box so that workers can handle radioactive materials in the box.

GPHS. General purpose heat source, consisting of eighteen RTGs to provide heat for spacecraft applications.

greenhouse effect. Heating of Earth's atmosphere caused by the presence of certain gases (e.g., carbon dioxide) that trap energy from sunlight striking the surface of Earth.

half-life. The length of time for a radioactive substance to lose half of its activity through radioactive decay. At the end of one half-life, only 50 percent of the original radioisotope remains.

high-level waste (HLW). Highly radioactive solid material that results from the chemical reprocessing of spent nuclear fuel.

hormesis. A term that describes biologically beneficial effects of low-level radiation.

HTO. Tritiated water; water consisting of one hydrogen atom, one tritium atom, and one oxygen atom.

hydrogen bomb. A very powerful explosive device deriving its energy from the fusion process.

IAEA. International Atomic Energy Agency located in Vienna, Austria.

infrared microspectroscopy (IMS). The study of the properties of material systems by means of their interaction with infrared radiation.

ion. An atom that has lost or gained one or more orbiting electrons, thus becoming electrically charged.

ion engine. An engine that generates thrust for a spaceship by ejecting accelerated ions out the back end of the spaceship.

ionization. Any process by which atoms, molecules, or ions gain or lose electrons.

ionizing radiation. Radiation capable of causing ionization of the atoms through which it travels.

irradiation. The result of ionizing radiation impinging upon a material.

isotopes. Atoms of the same element but with different mass numbers. They contain the same number of protons in their nucleus (hence, the same chemical properties), but different numbers of neutrons. Isotopes of carbon are C-12, C-13, and C-14.

KeV. A thousand electron volts.

labeling. The incorporation of a radioactive tracer into a molecular species or macroscopic sample where the emitted radiation facilitates detection.

linear hypothesis. The assumption that any radiation causes biological damage, according to a straight-line graph of adverse heath effects versus dose.

luminescence. Light emissions that cannot be attributed merely to the temperature of the emitting body, but result from electron bombardment.

mammogram. A radiographic examination of the breast to check for tumors.

MeV. Million electron volts.

monoclonal antibody. A highly specific antibody to which a radioisotope is attached so that the radiation can be deposited in the material to which the antibody seeks out and attaches.

mrem. Millirem; one-thousandth of a rem.

MRI. Magnetic resonance imaging; a medical diagnostic device to determine irregularities in the human anatomy. It works on the principle of a changing magnetic field to cause shifts in the nuclear spin of hydrogenous material.

NERVA. The Nuclear Engine for Rocket Vehicle Application program in which about twenty nuclear rocket engines were built and tested for eventual use in space vehicle propulsion.

neutron. A basic particle residing in the nucleus of an atom that is electrically neutral. It has a mass approximately equal to a proton.

neutron probes. Devices containing spontaneous neutron emitters that can be imbedded in an area where neutrons are needed for diagnostic purposes.

neutron slowing down. The process of neutrons colliding with matter and losing energy.

NMRI. Nuclear magnetic resonance imaging; this is identical to an MRI. The word *nuclear* was dropped for purely marketing reasons.

nonproliferation. A policy to discourage (and even prevent) nonnuclear weapons nations or terrorist groups from acquiring nuclear bombs.

NPT. Nuclear Nonproliferation Treaty; an international treaty ratified in 1970 in which signatory nations agreed to submit to international safeguards (administered by the IAEA) to prevent the spread of nuclear weapons.

nuclear rocket. A rocket engine powered by nuclear energy, usually utilizing hydrogen as the propellant.

nucleus. The central, positively charged, dense portion at the center of an atom. It contains protons and neutrons.

nuclide. A species of atom characterized by the number of protons, number of neutrons, and energy content in the nucleus.

pair-production. The conversion of a photon into an electron and a positron when the photon traverses the strong electric field near a nucleus.

periodic table. A table of the elements, written in sequence in the order of atomic number or atomic weight, arranged to illustrate similarities in the properties of the elements.

person-rems. A measurement of the total radiation dose received by a population. It is an average radiation dose in rems multiplied by the number of people in the population group.

PET. Positron emission tomography; a medical diagnostic device that uses ingested radioisotopes that emit positrons, which are detected by a coincidence counter at precisely opposite sides of the patient, to reveal bodily abnormalities.

photoelectric effect. A process wherein a photon is absorbed by an atom and an electron is ejected with the energy of the photon, less a binding energy.

photon. A quantum of electromagnetic radiation of energy $h\nu$ (h being Planck's constant and ν the frequency of the radiation).

pitchblende. A massive brown to black ore found in nature that contains uranium.

plasma. A completely ionized gas, composed entirely of a nearly equal number of positive and negative free charges (positive ions and electrons).

positron. A positively charged electron.

proton. A basic particle residing in the nucleus of an atom that is positively charged. It has a mass approximately equal to a neutron.

pulsed neutron source. A source of neutrons that is delivered in pulses. This is normally accomplished by an accelerator, where a beam of deuterons is periodically focused on a target of tritium to produce pulses of 14 MeV neutrons.

PWR. Pressurized water reactor; a light water-cooled nuclear reactor operated at high pressure without coolant boiling.

rad. A unit of radiation energy absorption. 100 rad = 1 gray.

radiation. Energy transported by particles or rays (waves) coming from radioactive substances or from certain machines.

radiation dispersal device (RDD). A conventional chemical explosive device impregnated with radioactive substances (sometimes referred to as a "dirty bomb").

radiation oncologist. A cancer medical physician who uses radiation for treatment.

radioactive decay. The process of a radioactive substance being transmuted into another substance, literally disappearing, via disintegration.

radioactivity. The spontaneous emission of radiation from the nucleus of an atom.

radiocardiogram. An x-ray recording of the variation with time of the concentration of a radioisotope in a heart chamber.

radiograph. The photographic image produced when x-rays or gamma rays are transmitted through an object.

radiography. The technique of transmitting x-rays, gamma rays, or neutrons through an object to determine any abnormalities.

radioisotope. An unstable isotope of an element that will eventually undergo radioactive decay.

radionuclide. A radioactive species of an atom characterized by the constitution of its nucleus.

radiotherapy. The utilization of radiation for medical therapy.

radiotracer. A radioactive isotope that, when attached to a chemically similar substance or injected into a biological or physical system, can be traced by radiation detection devices, permitting determination of the distribution or location of the substance to which it is attached.

radon. A radioactive gas produced by the decay of one of the daughters of radium.

radura. The flowerlike symbol that indicates food has been irradiated to assure the absence of any unwanted contamination.

reactor. A nuclear device involving neutrons to induce a fission chain reaction to generate heat.

recycling. The process of chemically processing spent nuclear fuel, extracting useable fuel and fabricating it into new fuel elements, and utilizing the fuel for subsequent reactor burn cycles.

rem. Roentgen equivalent man; a unit used in radiation protection to measure the amount of damage to human tissue from a dose of ionizing radiation. An average American receives about 0.370 rems of radiation per year.

RHU. Radioisotope heater units; nuclear "batteries" fueled by plutonium-238 to produce heat for spacecraft.

RIA. Radioimmunoassay; a medical diagnostic process to very accurately measure the concentration of biological materials in the blood (such as hormones, enzymes, and the hepatitis virus).

RIT. Radioimmunotherapy; a medial procedure for treating residual diseases such as cancer cells that may have been left behind after a lumpectomy.

RTG. Radioisotope thermoelectric generators; nuclear "batteries" fueled by plutonium-238 to produce electricity for spacecraft.

shielding. Materials, often concrete, water, or lead, placed around radioactive materials to protect personnel against the danger of strong radiation sources.

shipping cask. A container that provides appropriate shielding and structural rigidity for the transportation of radioactive material.

Sievert. The international unit used for measuring effective radiation dose. 100 Rem = 1 Sievert.

smart bullet. A popular term used for monoclonal antibodies.

SPECT. Single photon emission computed tomography; a medical diagnostic tool utilizing radioisotopes to determine bodily abnormalities. Tc-99m is commonly used in such machines, in which the gamma rays from this radioisotope are detected by radiation counters that rotate around the patient.

spent nuclear fuel (SNF). Nuclear fuel elements that are discharged from a nuclear reactor after they have been used to produce power.

spontaneous neutrons. Neutrons that are emitted from spontaneous fission. Californium-252 is a radioisotope that spontaneously fissions and produces neutrons during that process.

TAT. Targeted alpha therapy; the use of alpha particles as part of a monoclonal antibody therapeutic procedure.

teletherapy. Radiation treatment administered by using a source that is at a distance from the body, usually by employing gamma rays from radioactive sources or x-rays or protons from an accelerator.

thermal neutrons. Neutrons in thermal equilibrium with their surroundings. At room temperature, their mean energy is about 0.025 eV.

thermionic. A process whereby heat energy is directly converted into electricity. One electrode is heated to a high enough temperature to emit electrons and the other electrode (serving as an electron collector) is at a much lower temperature.

thermoelectric. A process whereby heat energy is directly converted into electricity utilizing the Seebeck effect.

TMI. Three Mile Island reactor; this reactor, located in Pennsylvania, suffered a core melt accident in 1979. Fortunately, no one was injured.

transmutation. The process of changing one isotope into another isotope through irradiation.

tritium. A radioisotope of hydrogen. Its nucleus contains one proton and two neutrons.

x-ray fluorescence (XRF). A process of irradiating an unknown substance, causing electrons in the unknown substance to move out of orbit, thereby releasing an x-ray characteristic of the unknown material.

x-rays. Electromagnetic radiation of energy greater than that of visible light, usually produced by an x-ray machine. Although x-rays and gamma rays are similar, x-rays emanate from outside the nucleus, whereas gamma rays emanate from within the nucleus.

NOTE: The author is indebted to the following references for much of the material used for these definitions:

1. Raymond L. Murray, *Understanding Radioactive Waste*, 4th ed. (Columbus, OH: Battelle Press, 1994).

2. Susan D. Wiltshire, *The Nuclear Waste Primer/The League of Women Voters Educational Fund*, rev. ed. (Washington, DC: OCRWM Information Center, 1993).

3. G. C. Lowenthal, and P. L. Airey, *Practical Applications of Radioactivity and Nuclear Radiations* (Cambridge: Cambridge University Press, 2001).

4. Sybil P. Parker, editor in chief, *Dictionary of Scientific and Technical Terms*, 4th ed. (New York: McGraw-Hill, 1989).

NOTES

CHAPTER 2. THRIVING IN RADIATION

1. Albert Reynolds, *Bluebells and Nuclear Energy* (Madison, WI: Cogito Books, 1996), pp. 29–30.

2. "Radiation and Life," Uranium Information Centre, Ltd., Melbourne, Australia, http://www.uic.com.au/ral.htm; US data added by author.

3. Alan E. Waltar, *America the Powerless: Facing Our Nuclear Energy Dilemma* (Madison, WI: Medical Physics Publishing, 1995), p. 71.

4. Bernard L. Cohen, *Before It's Too Late* (New York: Plenum, 1983), pp. 33–39.

5. Merril Eisenbud, *Environmental Radioactivity from Natural, Industrial, and Military Sources*, 3rd ed. (Orlando, FL: Academic Press, 1987).

6. Myron Pollycove, "Radiation-Induced versus Endogenous DNA Damage: Possible Effect of Inducible Protective Responses in Mitigating Endogenous Damage," *Human and Experimental Toxicology* 22 (2003): 290–306.

7. Ibid.

8. Ibid.

9. John Cameron, "What Does the Nuclear Shipyard Worker Study Tell Us?" *American Nuclear Society Winter Meeting* 71 (November 13, 1994): 36.

10. T. D. Luckey, *Hormesis with Ionizing Radiation* (Boca Raton, FL: CRC Press, 1980).

11. T. D. Luckey, *Radiation Hormesis* (Boca Raton, FL: CRC Press, 1991).

12. James B. Muckerheide, "Organizing and Applying the Extensive Data That Contradict the LNT," paper presented at the International Symposium on the Effects of Low and Very Low Doses of Ionizing Radiation on Human Health, World Council of Nuclear Workers, Versailles, France, June 1999.

CHAPTER 4. AGRICULTURE

1. "The Peaceful Atom," Uranium Information Centre, Ltd., Melbourne, Australia, 2003, http://www.uic.com.au/peac.htm.

2. *Induced Mutations and Molecular Techniques for Crop Production*, Proceedings of a Symposium, jointly organized by the IAEA and the FAO, June 19–23, 1995 (Vienna, Austria: International Atomic Energy Agency, 1995).

3. B. S. Ahloowalia, M. Maluszynski, and Karin Nichtertein, "Global Impact of Mutation-Derived Varieties," paper delivered to the Joint FAO/IAEA Division of Nuclear Techniques in Food and Agriculture, International Atomic Energy Agency, Vienna, Austria, February 2003, p. 2.

4. James Dargie, "Socio-Economic Benefits of Nuclear Applications in Food and Agriculture," paper delivered to the Joint FAO/IAEA Division of Nuclear Techniques in Food and Agriculture, Vienna, Austria, 2002, p. 6.

5. Brookhaven National Laboratory, "New Crops: Peppermint and Grapefruit," January 22, 2004, http://www.bnl.gov/bnlweb/history/fruits.html.

6. Dargie, "Socio-Economic Benefits of Nuclear Applications in Food and Agriculture," p. 4.

7. Jihui Qian and Alexander Rogov, "Atoms for Peace: Extending the Benefits of Nuclear Technologies," IAEA, 2003, http://www.iaea.or.at/worldatom/Periodicals/Bull371/qian.html.

8. "The Peaceful Atom."

9. Ibid.

10. Nuclear Energy Institute, "A Brief History of Food Irradiation," http://www.nei.org/index.asp?catnum=3&catid=176.

11. "Hamburger Recall Rises to 25 Million Pounds," CNN Interactive, U.S. News Story Page, August 21, 1997, http://www.cnn.com/US/9708/21/beef.update.

12. "Pennsylvania Firm Expands Recall of Turkey and Chicken Products for Possible Listeria Contamination," Food Safety and Inspection Service (FSIS), October 2002, http://www.fsis.usda.gov/oa/recalls/prelease/pr090-2202.htm.

13. Nuclear Energy Institute, "A Brief History of Food Irradiation."

14. Bhabha Atomic Research Centre, "Food Preservation by Radiation Processing," Food Technology Division, Mumbai-400085, India, http://www.barc.ernet.in/webpages/organzation/foodtd_home_page/rp.fl.html.

15. J. D. Corrigan, "Experiences in Selling Irradiated Foods at the Retail Level," in *Proceedings of IAEA Symposia on Cost-Benefit Aspects of Food Irradiation Processing*, Aix-en-Provence, France, March 1–5, 1993, pp. 447–54.

16. Reuters, "Food Industry Seen Embracing Irradiation," February 4, 2003; see also American Nuclear Society, "Irradiated Food, Good; Food-Borne Pathogens, Bad," *Nuclear News* (July 2003): 62.

17. Mohamed El-Baradei, open plenary presentation at the IAEA International Conference on Security of Radioactive Sources, Hofburg Palace, Vienna, Austria, March 10, 2003.

18. Raymond Durante, private communication with author, February 2, 2004.

19. Bhabha Atomic Research Centre, "Food Preservation by Radiation Processing."

20. Reuters, "Food Industry Seen Embracing Irradiation."

21. American Nuclear Society, "Irradiated Food, Good; Food-Borne Pathogens, Bad."

Chapter 5. Medicine

1. M. K. Murphy et al., "Evaluation of the New Cesium-131 Seed for Use in Low-Energy X-Ray Brachytherapy," *Medical Physics* 31, no. 6 (June 2004): 1529–38.

2. E. D. Pisano et al., "Human Breast Cancer Specimens: Diffraction-Enhanced Imaging with Histologic Correlation-Improved Conspicuity of Lesion Detail Compared with Digital Radiography," *Radiology* 214 (2000): 895–901; F. Arfelli et al., "Mammography with Synchrotron Radiation: Phase-Detection Techniques," *Radiology* 215 (2000): 286–93.

3. Glenn T. Seaborg and Henry N. Wagner, *Nuclear Medicine: 100 Years in the Making: 1896–1996* (Reston, VA: Society of Nuclear Medicine, 1996), p. 13.

4. James E. Udelson et al., "Myocardial Perfusion Imaging for Evaluation and Triage of Patients with Suspected Acute Cardiac Ischemia," *Journal of the American Medical Association* 288, no. 21 (December 4, 2002): 2693–2700.

5. E. Tanaka, "Instrumentation for PET and SPECT Studies," in *Tomography in Nuclear Medicine, Proceedings of IAEA Symposium* (Vienna, Austria: IAEA, 1995), pp. 19–29.

6. Seaborg and Wagner, *Nuclear Medicine*, p. 15.

7. Ibid., p. 16.

8. M. Lankosz et al., "Topographic and Quantitative Elemental Analysis of Human Central Nervous System Tissue," in *European Synchrotron Radiation Facility, Highlights 2002* (Grenoble, France: ESRF, 2003), pp. 87–88.

9. "Organ Irradiated Outside of Body, Then Reimplanted," *Nuclear News* (February 2003): 61.

10. Janice M. Horowitz, Alice Park, and Sova Song, "Anatomy of a Tumor," *Time,* February 18, 2002, pp. 54–58.

11. Lauran Neergaard, "Women Turn to New Breast Cancer Treatment," Yahoo! News, August 27, 2003, http://story.news.yahoo.com/news?tmpl=story&u=/ap/20030825/ap_.

12. "New Answer for Cancer?" *Oak Ridge National Laboratory Review* 36, no. 2 (2003): 26–27, www.ornl.gov/ORNLReview (interview with Rodger Martin).

13. T. E. Witzig et al., "Safety of Yttrium-90 Ibritumomab Tiuxetan Radioimmunotherapy for Relapsed Low-Grade, Follicular, or Transformed Non-Hodgkin's Lymphoma," *Clinical Oncology* 21, no.7 (April 1, 2003): 1263–70.

14. Sally DeNardo, "Radioimmunotherapy for Breast Cancer: Systemic Tumor-Targeted Irradiation," *Advances in Oncology* 15, no. 3 (1999): 23–29.

15. J. G. Jurcic et al., "Targeted Alpha Particle Immunotherapy for Myeloid Leukemia," *Blood* 15, no. 4 (August 2002): 1233–39.

16. Michael R. McDevitt et al., "Tumor Therapy with Targeted Atomic Nanogenerators," *Science* 294 (November 16, 2001): 1537–40.

CHAPTER 6. ELECTRICITY

1. Nuclear Energy Institute, "It's Easy Being Green," *Nuclear Energy Insight* (July 2003): 1.

2. Nuclear Energy Institute, "Nuclear Energy's Low Costs," *Nuclear Energy Insight* (June 2002): 6.

3. Energy Information Agency, US Department of Energy, August 8, 2003.

4. Alan E. Waltar, *America the Powerless: Facing Our Nuclear Energy Dilemma* (Madison, WI: Medical Physics Publishing, 1995).

5. "America's Energy Challenge: The Nuclear Answer," Symposium at the George Bush Presidential Conference Center, College Station, Texas, November 19, 2001.

6. Electric Power Research Institute, "Deterring Terrorism: Aircraft Crash Impact Analyses Demonstrate Nuclear Power Plant's Structural Strength," December 2002, http://www.nei.org/documents/EPRINuclearPlantStructureStudy2002/2.pdf.

7. Nuclear Energy Institute, "Nuclear Energy's Low Costs."

8. Ann Bisconti, "Why Public Support for Nuclear Energy Has Increased," *Natural Gas* (Wiley Periodicals) 19, no. 10 (May 2003): 15–21.

9. Ibid.

10. Waltar, *America the Powerless*, p. 3.

11. S. R. Hatcher, ed., and M. Hori, committee chair, International Nuclear Societies Council, *A Vision for the Second Fifty Years of Nuclear Energy* (Chicago: American Nuclear Society, 1996).

12. Scott Heaberlin, *A Case for Nuclear-Generated Electricity* (Columbus, OH: Battelle Press, 2003).

CHAPTER 7. MODERN INDUSTRY

1. Roger Besdek, *The Untold Story: The Economic Benefits of Nuclear Technologies* (Washington, DC: Management Information Services, 1996).

2. G. C. Lowenthal and P. L. Airey, *Practical Applications of Radioactivity and Nuclear Radiations* (Cambridge: Cambridge University Press, 2001). A substantial amount of the material included in this chapter was obtained from this excellent book.

3. Paul Frame and William Kolb, *Living with Radiation: The First Hundred Years* (Edgewater, MD: Syntec Inc., 2002).

4. International Atomic Energy Agency, "Building a Better Future: Contributions of Nuclear Science and Technology," IAEA/PI/A58E/98-00206, February 1998, p. 37.

5. Lowenthal and Airey, *Practical Applications of Radioactivity and Nuclear Radiations*, p. 191.

6. Lowenthal and Airey, *Practical Applications of Radioactivity and Nuclear Radiations*, p. 198.

7. Ibid., p. 200.

8. International Atomic Energy Agency, "Building a Better Future," p. 43.

9. International Atomic Energy Agency, *Radioisotope Applications for Troubleshooting and Optimizing Industrial Processes* (Vienna, Austria: International Atomic Energy Agency, March 2002), p. 9.

Chapter 8. Transportation

1. Dick Kovan, "Selling Neutrons and X-Rays to Engineers," American Nuclear Society, *Nuclear News* (December 2003): 34.

2. P. J. Webster et al., "Residual Stresses in Railway Rails—The FaME38 Project," in *European Synchrotron Radiation Facility, Highlights 2002* (Grenoble, France: ESRF, February 2003), p. 92.

3. Raymond Murray, *Introduction to Nuclear Engineering*, 2nd ed. (Englewood Cliffs, NJ: Prentice-Hall, 1961), p. 339.

4. John W. Simpson, *Nuclear Power from Underseas to Outer Space* (La Grange Park, IL: American Nuclear Society, 1995), p. 73.

5. Murray, *Introduction to Nuclear Engineering*, p. 342.

Chapter 9. Space Exploration

1. US Department of Energy, Office of Nuclear Energy, Science & Technology, "Nuclear Power in Space," DOE/NE-0071, http://www.nuc.umr.edu/nuclear_facts/spacepower/spacepower.html.

2. George Shortley and Dudley Williams, *Elements of Physics*, 2nd ed. (Englewood Cliffs, NJ: Prentice-Hall, 1955), p. 663.

3. US Department of Energy, Office of Nuclear Energy, Science & Technology, "Nuclear Power in Space."

4. Ibid.

5. Ibid.

6. http//:spaceprojects.arc.nasa.gov/Space_Projects/pioneer/PNhome.html.

7. Highlights, "Voyager Now: The End Is Nigh," *Nature* 426 (November 6, 2003): supplementary articles on pp. 21–22, 45–48, 48–51.

8. US Department of Energy, Office of Nuclear Energy, Science & Technology, "Nuclear Power in Space."

9. Regina Hagen, "Nuclear Powered Space Missions—Past and Future," http://www.globenet.free-online.co.uk/ianus/npsm2.htm.

10. "Cassini," http://nssdc.gsfc.nasa.gov/planetary/cassini.html.

11. Leonard David, "Space Nuclear Power Viewed as 'Must Have' Technology," February 2, 2003, http://www.space.com/businesstechnology/technology/staif_day1_030204.html.

12. "Nuclear Rocket Propulsion—The 1960s—Nuclear Thermal Propulsion," http://www.fas.org/nuke/space/c04rover.htm.

13. Greg Clark, "Will Nuclear Power Put Humans on Mars?" http://www.space.com/scienceastronomy/solarsystem/nuclearmars_000521.html.

14. David Whitehouse, "NASA to Go Nuclear," Science/Nature, http://news.bbc.co.uk/1/hi/sci/tech/2684329.stm.

15. Clark, "Will Nuclear Power Put Humans on Mars?"

16. Hagen, "Nuclear Powered Space Missions—Past and Future."

17. David Boyle, Department of Nuclear Engineering, Texas A&M University, private communication with author, April 2000.

CHAPTER 10. TERRORISM, CRIME, AND PUBLIC SAFETY

1. Katherine Ramsland, "Hair Evidence Analysis," http://www.crimelibrary.com/criminal_mind/forensics/trace/5.html?sect=21.

2. Hendrik Ball, "Arsenic Poisoning and Napoleon's Death, http://www.victorianweb.org/history/arsenic.html.

3. Crime Library, "Evidence for Murder," http://www.crimelibrary.com/terrorists_spies/assassins/napoleon_bonaparte/5.html?sect=24.

4. Jeffrey Kluger, "How Science Solves Crimes," *Time*, October 21, 2002, p. 39.

5. Donald Bowling, "The Use of DNA Fingerprinting in Forensic Science," http://www.esva.net/~dbowling/Dna.htm.

6. John Medina, *The Outer Limits of Life* (Nashville: Oliver Nelson, a Division of Thomas Nelson Publishers, 1991), p. 23.

7. Ibid.

8. Bowling, "The Use of DNA Fingerprinting in Forensic Science."

9. Jon Cohen, "Genes and Behavior Make an Appearance in the OJ Trial," *Science* 268 (April 7, 1995): 22–23.

10. Bowling, "The Use of DNA Fingerprinting in Forensic Science."

11. William Illsey Atkinson, "Infrared Microspectroscopy Proves to Be a Powerful Detection Tool," http://pubs.acs.org/hotartcl/tcaw/00/oct/atkinson.html.

12. Ibid.; T. J. Wilkinson et al., "Physics and Forensics," *Physics Today* (March 2002): 43–46; Jenny Duong, "Berkeley Lab Scientists Fight Crime with Laser, *Daily Californian*, October 19, 2003, http://www.dailycal.org/article.asp?id=9465.

13. Duong, "Berkeley Lab Scientists Fight Crime with Laser."

14. Illsey Atkinson, "Infrared Microspectroscopy Proves to Be a Powerful Detection Tool."

15. Paul Eng, "(Neutron) Bomb Sniffer: New Sensor to Detect Bombs and Contraband in Luggage," ABC News, http://i.abcnews.com/sections/scitech/FutureTech/futuretech030204.html.

16. Arden D. Dougan, Lawrence Livermore National Laboratory, "The Safe Cargo Challenge," talk given to the American Nuclear Society, Northern California section, Fremont, California, March 6, 2003.

17. Paul Pettit et al., for Vehicle and Cargo Inspections System (VACIS), "Characterization Technologies," chap. 19 in *Nuclear Facilities Decommissioning Handbook*, ed. Anibal L. Tabois, Thomas L. LaGuardia, and A. Alan Moghissi (Boca Raton, FL: CRC Press, in publication).

18. Bruce L. Freeman, "Plasma Focus Research—Reasons for Continuing Efforts," *Current Trends in International Fusion Research—Proceedings of the Fourth Symposium*, ed. Charles D. Orth, Emillio Panarella, and Richard F. Post (Ottawa: NRC Research Press, National Research Council of Canada, 2001).

19. Ralph James, "Development of a Field-Portable, X-Ray Spectrometer System for Detecting Landmines," Sandia National Laboratories, Rbjames@Sandia.Gov.

20. J. A. Bamberger et al., "Timed Neutron Detection for Land Mines," *2003 IEEE Nuclear Science Symposium and Medical Imaging Conference*, Portland, Oregon, October 16–26, 2003 (Portland: NSSS/MIC, 2003).

21. Nuclear News Interview, "Desrosiers: Irradiation in Homeland Security," American Nuclear Society, *Nuclear News* (December 2003): 29–32.

22. Ibid.

23. Bernard L. Cohen and I. S. Lee, "A Catalog of Risks," *Health Physics* 36, no. 707 (1979); see also the update, Bernard L. Cohen, "Catalog of Risks Extended and Updated," *Health Physics* 61, no. 3 (September 1991): 317–35.

24. Alan E. Waltar, *America the Powerless: Facing Our Nuclear Dilemma* (Madison, WI: Medical Physics Publishing, 1995), chap. 8, pp. 141–60.

25. Garret Condon, "World Is Scary, But Terrorism Less of a Threat Than Obesity," (Kennewick, Washington) *Tri-City Herald*, April 17, 2003, p. D3.

Chapter 11. Arts and Sciences

1. James C. Chatters, "Kennewick Man, Northern Clans, Northern Traces," http://www.mnh.si.edu/arctic/html/kennewick_man.html.

2. "An Accidental Discovery," http://www.washington.edu/burkemuseum/kman/virtualexhibit_intro.htm.

3. "Ancient People in the Americas," http://www.washington.edu/burkemuseum/kman/ancientpeoples.htm.

4. Thomas Higman, "The 14Carbon Method," http://www.c14dating.com/int.html.

5. Ibid.

6. Raymond Murray, *Nuclear Energy: An Introduction to the Concepts, Systems, and Applications of Nuclear Processes*, 5th ed. (Boston: Butterworth-Heinemann, 2001), p. 231.

7. Amos Frumkin, "Radiometric Dating of the Siloam Tunnel, Jerusalem," *Nature* 425 (September 11, 2003): 169–71.

8. Murray, *Nuclear Energy*, p. 231.

9. T. C. Partridge, Darryl Granger, and R. J. Clarke, "Lower Pliocene Hominid Remains from Sterkfontein," *Science* 300 (April 25, 2003): 607–12.

10. David Filmore, "An Aging Question," *Today's Chemist at Work* (January 2004): 39.

11. R. E. Taylor and Martin J. Aitken, eds., *Chronometric Dating in Archaeology* (New York: Plenum, 1997).

12. Murray, *Nuclear Energy*, p. 235.

13. "Fraud and Misrepresentation in Colored Gems," http://www.gemshopper.com/fraud.html.

14. "Art and Stone," *CLEFS, Commissariat a l' energie atomique (CEA)* 34 (Winter 1996–1997): 49.

15. *Articulations at About*, transcript, March 18, 2001, Speaker J. Barto Arnold, http://archeology.abouit.com/library/chat/n_arnold.htm.

CHAPTER 12. ENVIRONMENTAL PROTECTION

1. International Atomic Energy Agency, "Building a Better Future: Contributions of Nuclear Science and Technology," IAEA/PI/A58E, February 1998, p. 41.

2. International Atomic Energy Agency, "Building a Sustainable Future," IAEA/PI/B06E-Rev.1/02-01254, June 2002, p. 8.

3. International Atomic Energy Agency, "Isotopes in Water and Environmental Management," 95-02889 IAEA/PI/A44E, September 1995, p. 5.

4. Ibid.

5. Peter Rickwood, "Water for Development: World Water 2002 Points to Mounting Challenges," *IAEA Bulletin* 44, no. 1 (June 2002): 22.

6. Ibid., p. 21.

7. Ibid.

8. Ibid.

9. G. C. Lowenthal and P. L. Airey, *Practical Applications of Radioactivity and Nuclear Radiations* (Cambridge: Cambridge University Press, 2001), p. 279.

10. Ibid., p. 278.

11. Ibid., p. 298

12. Ibid., p. 299.

13. Ibid., p. 300.

14. International Atomic Energy Agency, "Building a Better Future: Contributions of Nuclear Science and Technology," p. 47.

15. Ibid.

16. Nuclear Energy Institute, "Cool, Clean Water: Nuclear Desalination Provides Hope for Developing World," *Nuclear Energy Insight* (February 2003): 7.

17. Ibid.

18. International Atomic Energy Agency, "Building a Better Future: Contributions of Nuclear Science and Technology," p. 47.

19. Lowenthal and Airey, *Practical Applications of Radioactivity and Nuclear Radiations*, p. 301.

20. International Atomic Energy Agency, "Building a Better Future: Contributions of Nuclear Science and Technology," p. 48.

21. International Atomic Energy Agency, "Science Serving People," IAEA/TC, 02-01407, July 2002, pp. 36–39.

22. Ibid., pp. 34–35.

23. Brad Lemley, "The New Ice Age," *Discover Magazine* 23, no. 9 (September 2002): 35–47.

24. Lowenthal and Airey, *Practical Applications of Radioactivity and Nuclear Radiations*, p. 297.

25. Ibid., p. 289.

26. Ibid.

27. Centers for Disease Control and Prevention (CDC) and Fred Hutchinson Center Research Center, *A Guide to the Hanford Thyroid Disease Study—Final Report* (Atlanta: CDC, June 21, 2002); John Till et al., *A Risk-Based Screening Analysis for Radionuclides Released to the Columbia River from Past Activities at the U.S. Department of Energy Weapons Site in Hanford, Washington* (Neeses, SC: Risk Assessment Corporation, 2002).

28. Roy E. Gephart, *Hanford: A Conversation about Nuclear Waste and Cleanup* (Columbus, OH: Battelle Press, 2003).

29. Don J. Bradley, *Behind the Nuclear Curtain: Radioactive Waste Management in the Former Soviet Union* (Columbus, OH: Battelle Press, 1997).

30. IAEA-MEL (Marine Environment Laboratory), "Programmes Related to the Radioactive Waste Dumped in the Arctic Seas," Report of Recent Activities, IAEA/PI/A43, 94-3467 (Monaco: International Atomic Energy Agency, 1994), p. 24.

31. Raymond L. Murray, *Understanding Radioactive Waste*, 4th ed. (Columbus, OH: Battelle Press, 1994), p. 103.

32. J. Lovell, "Asian Smog Cloud Threatens Millions, Says U.N.," Reuters, August 11, 2002.

33. United Nations Environmental Assessment Report, "Asian Brown Cloud: Climate and Other Environmental Impacts," August 14, 2004, http://www.rrcap.unep.org/issues/air/impactstudy/index.cfm; also presented as "Asian Brown Cloud Heating Europe," *London Times Online*, August 12, 2002.

34. International Atomic Energy Agency, "Building a Better Future: Contributions of Nuclear Science and Technology," p. 43.

35. David A. Scott, University of Victoria, Victoria, British Columbia, Canada, private communication with author during American Nuclear Society winter meeting, November 12, 2001.

36. Hans-Holger Rogner, International Atomic Energy Agency, Vienna, Austria, private communication with author during American Nuclear Society winter meeting, November 12, 2001.

37. Alan E. Waltar, *America the Powerless: Facing Our Nuclear Energy Dilemma* (Madison, WI: Medical Physics Publishing, 1995), p. 4.

38. Joseph V. Spadaro, Lucille Langlois, and Bruce Hamilton. "Assessing the Difference: Greenhouse Gas Emissions of Electricity Generating Chains," *IAEA Bulletin* 42, no. 2 (February 2000): 21.

39. Bruno Comby, *Environmentalists for Nuclear Energy* (Paris: TNR Editions, 2001).

CHAPTER 13. MODERN ECONOMY

1. Management Information Services, *The Untold Story: Economic and Employment Benefits of the Use of Radioactive Materials* (Washington, DC: Management Information Services, Inc., 1994); Management Information Services, *Economic and Employment Benefits of the Use of Nuclear Energy to Produce Electricity* (Washington, DC: Management Information Services, Inc., 1994).

2. Management Information Services, *The Untold Story: The Economic Benefits of Nuclear Technologies* (Washington, DC: Management Information Services, Inc., 1996).

3. Ibid.

4. Ibid.

5. Kazuaki Yanagisawa et al., "An Economic Index Regarding Market Creation of Products Obtained from Utilization of Radiation and Nuclear Energy (IV), Comparison between Japan and the U.S.A.," *Journal of Nuclear Science and Technology* 39, no. 10 (October 2002): 1120–24; Seiichi Tagawa et al., "Economic Scale of Utilization of Radiation (I): Industry, Comparison between Japan and the U.S.A.," *Journal of Nuclear Science and Technology* 39, no. 9 (September 2002): 1002–1007; T. Kume et al., "Economic Scale of Utilization of Radiation (II): Agriculture," *Journal of Nuclear Science and Technology* 39, no. 10 (2002): 1106; T. Unoue et al., "Economic Scale of Utilization of Radiation (III): Medicine," *Journal of Nuclear Science and Technology* 39, no. 10 (2002): 1114.

6. Tagawa et al., "Economic Scale of Utilization of Radiation (I): Industry Comparison between Japan and the U.S.A."

7. Zhu Szhuzhong, Private communication with author, July 1, 2002; Chen Dianhua, "Present Situation and Development of Isotope and Radiation Processing in China," China Isotope & Radiation Association, transmitted by Zhu Szhuzhong, private communication with author, July 1, 2002.

Chapter 15. A Glimpse into the Future

1. Avraham Dilmanian, "Microbeam Radiation Therapy," *Battelle World*, April 21, 2003.

2. European Synchrotron Radiation Facility, *European Synchrotron Radiation Facility, Highlights 2002* (Grenoble, France: ESRF, 2002), pp. 87–88.

3. International Atomic Energy Agency, "Building a Better Future: Contributions of Nuclear Science and Technology," IAEA/PI/A58E, 98-00206, February 1998, p. 37.

4. "Organ Irradiated Outside of Body, Then Reimplanted," *Nuclear News* (February 2003): 61.

5. Jerry M. Cuttler, Myron Pollycove, and James S. Welsh, "Application of Low Doses of Radiation for Curing Cancer," paper presented at the ANS/ENS Winter Meeting, Washington, DC, November 15, 2000; see also http://cnts.wpi.edu/RHS/Docs/cuttler_et_al.htm.

6. Alan E. Waltar and Albert B. Reynolds, *Fast Breeder Reactors* (Elmsford, NY: Pergamon Press, 1981).

7. "What Is ITER?" January 2004, http://www.iter.org/ITERPublic/ITER/whatis.html.

8. Ron Lehman et al., *Atoms for Peace after 50 Years: The New Challenges and Opportunities* (Washington, DC: Center for Global Security Research [GCSR], Lawrence Livermore National Laboratory, December 2003), p. B-34.

9. Stephen R. Covey, *The 7 Habits of Highly Effective People* (New York: Simon & Schuster, 1989), p. 95.

10. Chauncey Starr, "National Energy Planning for the Century: The Continental SuperGrid," *Nuclear News* (February 2002): 31–35.

11. "Reaching for the Stars, Antimatter," April 12, 1999, http://science.nasa.gov/newhome/headlines/prop12apr99_1.htm.

INDEX

NOTE: Page numbers in **bold** indicate the word's location in the glossary